Virology Research Progress Series

INSECT VIRUSES: DETECTION, CHARACTERIZATION AND ROLES

Virology Research Progress Series

Insect Viruses: Detection, Characterization and Roles
Christopher I. Connell and Dominick P. Ralston (Editors)
2009. ISBN: 978-1-60692-965-0

Virology Research Progress Series

INSECT VIRUSES: DETECTION, CHARACTERIZATION AND ROLES

**CHRISTOPHER I. CONNELL AND
DOMINICK P. RALSTON
EDITORS**

Nova Biomedical Books
New York

Copyright © 2009 by Nova Science Publishers, Inc.

All rights reserved. No part of this book may be reproduced, stored in a retrieval system or transmitted in any form or by any means: electronic, electrostatic, magnetic, tape, mechanical photocopying, recording or otherwise without the written permission of the Publisher.

For permission to use material from this book please contact us:
Telephone 631-231-7269; Fax 631-231-8175
Web Site: http://www.novapublishers.com

NOTICE TO THE READER

The Publisher has taken reasonable care in the preparation of this book, but makes no expressed or implied warranty of any kind and assumes no responsibility for any errors or omissions. No liability is assumed for incidental or consequential damages in connection with or arising out of information contained in this book. The Publisher shall not be liable for any special, consequential, or exemplary damages resulting, in whole or in part, from the readers' use of, or reliance upon, this material. Any parts of this book based on government reports are so indicated and copyright is claimed for those parts to the extent applicable to compilations of such works.

Independent verification should be sought for any data, advice or recommendations contained in this book. In addition, no responsibility is assumed by the publisher for any injury and/or damage to persons or property arising from any methods, products, instructions, ideas or otherwise contained in this publication.

This publication is designed to provide accurate and authoritative information with regard to the subject matter covered herein. It is sold with the clear understanding that the Publisher is not engaged in rendering legal or any other professional services. If legal or any other expert assistance is required, the services of a competent person should be sought. FROM A DECLARATION OF PARTICIPANTS JOINTLY ADOPTED BY A COMMITTEE OF THE AMERICAN BAR ASSOCIATION AND A COMMITTEE OF PUBLISHERS.

Library of Congress Cataloging-in-Publication Data

Connell, Christopher J.
  Insect viruses : detection, characterization, and roles / Christopher J. Connell and Dominick P. Ralston.
       p. cm. -- (Virology research progress)
  Includes bibliographical references and index.
  ISBN 978-1-60692-965-0 (hardcover : alk. paper)
  1. Insects--Viruses. 2. Medical virology. 3. Viral pesticides. I. Ralston, Dominick P. II. Title.
  QR327.C66 2009
  616.9'101--dc22
                                                            2009009165

Published by Nova Science Publishers, Inc. ✢ *New York*

Contents

Preface vii

Research and Review Studies

Chapter 1	Insect Virus Proteins Involved in the Peroral Infectivity of the Viruses and Their Potential Practical Application in Pest Control *Wataru Mitsuhashi*	1
Chapter 2	RNAi and the Study of Insect Immunity *E. P. Silva and N. A. Ratcliffe*	21
Chapter 3	Epidemiology of Kakugo Virus, An Insect Picorna-like Virus Identified in Aggressive Honeybee Workers – A Study of Virus-host Interactions in the Honeybee Society *Tomoko Fujiyuki*	39
Chapter 4	Honeybee Viruses in Uruguay: First Detection of Honeybee Viruses in South America and Their Potential Role in Mortality of Honeybees *Karina Antúnez, Bruno D'Alessandro and Pablo Zunino*	57
Chapter 5	Encephalitic Arboviruses: Emerging and Re-Emerging Problem *Agostino Pugliese, Tiziana Beltramo and Donato Torre*	75
Chapter 6	Neurologic Manifestations of West Nile Virus Infection *Ronen Spiegel and Yoseph Horovitz*	101
Chapter 7	Effects of Coinfection with Borrelia Burgdorferi and Anaplasma Phagocytophilum in Vector Ticks and Vertebrate Hosts *Michael L. Levin*	119

Short Communications

Early Detection of Baculovirus Expression and Infection in Lepidopteran Larvae Fed Occlusion Bodies of an AcMNPV Recombinant Carrying a Red Fluorescent Protein Gene **147**
Arthur H. McIntosh and James J. Grasela

Reovirus-like Double Stranded RNA Fractions in a *Drosophila Melanogaster* Line Containing Individual Second Chromosome from Natural Population **157**
E. G. Pasyukova and D. V. Mukha

Cellular Secretion of Sf21 Cells upon Baculovirus Infection **165**
Xiao-Wen Cheng

Index **169**

Preface

Insects are a major group of arthropods and the most diverse group of animals on the earth, with over a million described species. In common with all other life forms, insects suffer from viruses that cripple and kill. Admittedly, insects transmit viruses that cause illness in humans, but the insect in such cases is usually unaffected. This book includes research on such viruses affecting humans. Also included in this book are the ways to recognize insect viruses and their use in pest control. Basic and applied research on insect virus proteins associated with the peroral infectivity is reviewed as well. Furthermore, RNAi has been applied to a number of studies involved in insect immunity. Thus, the effect of RNA interference on viral infections is also studied.

Chapter 1 - The midgut is the initial place of infection of host insects by insect viruses that are perorally acquired. This initial (primary) infection leads to their replication in the main target tissues for replication in parts of the insects. Many studies have identified the range of proteins in many insect viruses, which are involved in their initial infection of the midgut. Some proteins are essential for peroral infection, and others only enhance the infection; some of the proteins initiate or enhance the infections by working the initial-infection tissue, called the midgut epithelium, and the others facilitate passage of the viruses across an acellular membrane, the peritrophic membrane (which is a barrier against the initial infection) toward the midgut epithelium by disrupting the membrane. Many studies have suggested that this enhancement of infectivity by some of the proteins is potentially applicable to pest control in agriculture and forestry, because the proteins are potential synergists for insect viral insecticides, and the co-administration of the proteins will drastically increase the efficacy of the viral pesticides. This will encourage far wider use of viral pesticides that have an advantage of being safer for vertebrates, plants, and the environment than synthetic pesticides. This chapter reviews basic and applied research on insect virus proteins associated with peroral infectivity, especially with respect to entomopoxviruses and granuloviruses, which belong to the families *Poxviridae* and *Baculoviridae*, respectively.

Chapter 2 - RNA interference is a post-transcriptional mode of gene silencing that works as a cellular response to gene invasion resulting from transposons or viruses. Silencing occurs by a double-stranded RNA (homologous to the silenced gene) targeting and degrading the mRNA. To date, RNAi has been applied to a number of studies involved in insect immunity, such as the functional characterization of antimicrobial peptides and pattern recognition proteins, the encapsulation and phagocytosis processes, and the effects of viral infections and the modulation of the host immune response by insect parasitoids. These studies are reviewed in the present chapter and the importance of the application, as well as the usefulness of RNAi as a tool, are discussed in the light of the results obtained so far from studies on insect immunity.

Chapter 3 - Honeybee (*Apis mellifera* L.) is a eusocial insect. Honeybees work together to create a nest (colony), obtain food, and maintain their shelter. In their colonies, female adult bees divide the various labors, such as reproduction and colony maintenance, to propagate effectively as a super-organism. Sociality and group living are not always beneficial, however, because animals in groups tend to attract predators. In addition, once an infectious disease occurs, it prevails in the group rather easily because of the dense population. Therefore, social animals may have specific defense systems for protection against predators and pathogens.

We have identified a novel insect picorna-like virus, Kakugo virus, from the brains of aggressive honeybees; that is, honeybees that counterattack their natural enemy, and studied its epidemiology. In this review, I discuss the interaction between Kakugo virus and the host honeybee, and briefly describe other insect picorna-like viruses.

Chapter 4 - More than 18 RNA viruses that affect honeybees have been reported. Most of them cause unapparent infections or persist in a latent form in healthy colonies, while only a few viruses produce clinical symptoms easily recognizable by beekeepers. The presence of these viruses has been associated with honeybee mortality episodes that have been taking place during the last years worldwide. Several reports have been recently published in North America and Europe, although precise information about bees' mortality and the presence of viruses in South America was lacking.

In the present work, we report the presence of different RNA viruses, including acute bee paralysis virus, chronic bee paralysis virus, black queen cell virus, sacbrood virus and deformed wing virus in honeybee samples from different locations in Uruguay. This was the first report about the presence of potentially pathogenic honeybee viruses in South America, and their relation with other honeybee pathogens such as *Varroa destructor* and *Nosema* spp. is discussed.

The detection of viruses in different geographic regions in the country, the simultaneous co-infection of colonies by several viruses and even other pathogens, and the fact that in most of the samples one or more viruses were found, indicate that they are widely spread in the region.

Chapter 5 - At present Arboviruses include 534 viruses of them 134 are of human interest, and a large part of the last ones are endowed with neurotropism. In particular, this ecological classification concerns different families and viral genera, but all the viruses included in the same classification, are characterized by transmission through arthropod bites. These viruses can be responsible for human infections ranging from asymptomatic or mild ones to fatal illness, like encephalitis and hemorrhagic fevers. The most important arboviruses causing human pathology are included in three viral families, Togaviridae (genus Alphavirus), Flaviviridae, and Bunyaviridae.

Different distribution of encephalitic arboviruses exists among various areas of the world, however generally these viruses are present, in large or small degree, in all the temperate or warm climate regions, and sometimes also in more cold areas (but in this case only in the hot season). The diffusion of these infections parallels the cycle of their vectors: especially mosquitoes, ticks and phlebotomes. Habitat modification induced by men, as deforestation, can change the ecology of vectors and the epidemiology of encephalitides transmitted by arboviruses, but also of other vector–borne diseases. This phenomenon is due both to climate and to demographic changes, and in particular occurs in South America, West and Central Africa and in South-East Asia. Moreover also the rapidity of modern transports has contributed to encephalitic arboviruses new localization areas that sometimes are more adapted to cause major epidemic phenomena. The modification in reservoir host environment and the viral agents adaptation to new hosts, represent the main cause of pattern modification of these infections in the world. West Nile virus is a classic example of this phenomenon. In fact, recently, this virus was introduced in the Western Hemisphere. Consequently arboviruses can no more be considered specific of particular regions of the world, but represent a more wide potential danger for men. Since the acquired immunity of population against some arboviruses and the lability in the external environment of a large part of them may risk their extinction, the evolution has adopted a dual host tropism strategy for viral survival: reservoir hosts and final, or occasional hosts. Men just sometimes represent the last ones that may have clinical manifestations.

In order to prevent these serious pathologies, because there are no specific therapies for arboviral encephalitides, fight against vectors and vaccines employ can constitute the only defences for these diseases. For example effective vaccines exist against Tick-borne encephalitis virus (TBEV) and Japanese encephalitis virus (JEV).

In conclusion in our chapter we analyse the general questions concerning the arboviral encephalitides in the world (biological characteristics of the causative agents, epidemiology, pathogenicity, diagnosis and prophylaxis) and in the last part of this paper we analyse the Italian situation. In fact, beginning from 2000 we have studied the epidemiology of arboviruses in Piedmont and in particular in the Turin Province. Finally we report also some data concerning arboviral biology.

Agostino Pugliese is Professor of Clinical Microbiology at the Medical Faculty of Turin University. Dr. Donato Torre is Specialist of Infectious Diseases, and of Paediatrics.

Chapter 6 - West Nile virus (WNV) infection is a zoonotic disease. The virus belongs to the family Flaviviridae. Wild birds serve as the reservoir of the virus in nature and mosquitoes mainly from the culex species serve as the vector for its replication cycle.

Humans as well as horses and other domestic animals are incidental hosts.

Until recently, WNV infections were traditionally considered to be a mild self-limited disease. When symptomatic it usually manifested with fever, headache, myalgia, generalized lymphadenopathy and rash. Central nervous system (CNS) involvement was infrequent. In the last decade, several epidemics of WNV have occurred in several areas in Europe, in Israel and, for the first time, in the Western Hemisphere, accounting for a large outbreak in New York. These epidemics were characterized by an exceptionally high rate of neurological morbidity and mortality.

The most common neurological involvement are encephalitis/meningoencephalitis followed by aseptic meningitis and flaccid paralysis. Other less common neurological presentations include Guillain-Barre syndrome, rhombencephalitis, optic neuritis and aphasia. The cause for this clinical evolution from a mild flu-like illness to a severe neuroinvasive disease is not fully understood and is thought in part to be related to the emergence of more neurovirulent WNV strains.

Chapter 7 - Agents of Lyme disease - *Borrelia burgdorferi* and human granulocytic anaplasmosis - *Anaplasma phagocytophilum* (formerly *Anaplasma phagocytophila*) are perpetuated in natural cycles involving the black-legged tick (*Ixodes scapularis*) and its vertebrate hosts. This predetermines the exposure of humans as well as wild and domestic animals to both pathogens and consequent concurrent infections. We studied whether a preexisting infection with either *Borrelia* or *Anaplasma* would affect acquisition and transmission of a second pathogen in ticks and their mammalian hosts. Also, we assessed the efficiency by which individual nymphs could transmit either agent alone or both agents simultaneously, to individual susceptible hosts. There was no evidence of interaction between the agents of Lyme disease and human granulocytic anaplasmosis in *I. scapularis* ticks. The presence of either agent in ticks did not affect acquisition of the other agent from an infected host. Transmission of the agents of Lyme disease and human granulocytic anaplasmosis by individual ticks was equally efficient and independent. Dually infected ticks transmitted each pathogen to susceptible hosts as efficiently as ticks infected with one pathogen only. On the other hand, a primary infection with either *B. burgdorferi* or *A. phagocytophilum* in mice inhibited acquisition and transmission of a second agent, suggesting interference between these two agents in a vertebrate host. Consequences of co-infection in ticks, wild animals and humans are discussed.

Short Communication A - A method was devised utilizing a baculovirus recombinant (AcMNPV hsp70Red) carrying a red fluorescent protein (RFP) gene under the early heat shock promoter (hsp70) to assess potential infectivity of larvae fed occlusion bodies. A time study was employed whereby first and third instars of *Trichoplusia ni*, *Heliothis subflexa* and *Helicoverpa zea* could be followed for red fluorescence under a UV-light inverted microscope at various time intervals following diet surface feeding of recombinant occlusion bodies. Larvae of *T. ni* and *H. subflexa* that are permissive to AcMNPV showed 100 % fluorescence by 24h and 36h respectively for 1st instars exposed to 1×10^5 OB. Even when the concentration of OB was reduced 100 fold to 1×10^3 OB, 94%-100% fluorescence was observed by 24h post feeding. *H. zea* which is less permissive to AcMNPV than *T. ni* and *H. subflexa* and when 1st instars were fed 1×10^5 OB showed 100% red fluorescence at 48h post exposure as contrasted with 86%-90% fluorescence at the 96h exposure period for the lower dose of 1×10^3 OB. *T. ni* and *H. subflexa* third instars fed 1×10^5 OB gave 100% fluorescence between 24h (*T.ni*) and 36h- 72h (*H. subflexa*) post exposure whereas at the lower dose of 1×10^3 OB 100% fluorescence was achieved between 24h-72h post exposure. Third instar *H. zea* larvae fed both doses did not attain 100% fluorescence. All larvae of each species that gave 100% fluorescence resulted in 100% mortality. Fluorescent fifty (FL_{50}) values, the time taken to attain 50% fluorescence of exposed larvae, were calculated for all categories. *T. ni* gave the shortest times for both 1st (FL_{50} = 12.8h) and 3rd instar (FL_{50} =

16.0h) at the OB dose of 1×10^5 followed by *H. subflexa* and *H. zea*. The same pattern was observed at the lower OB dose of 1×10^3.

Short Communication B - DNA samples extracted from 20 *Drosophila melanogaster* lines containing individual second chromosomes from Raleigh population in the genetic background of the isogenic laboratory line Samarkand (Samarkand;Raleigh$_i$;Samarkand) were analyzed using electrophoresis in agarose gel. In two samples, in addition to high molecular weight genomic DNA, ten separate bands of nucleic acid material were found. One of the two lines was analyzed in more detail. All additional bands were resistant to DNAse and RNAse A treatment but sensitive to RNAse III treatment. The size of the bands varied approximately from 1 to 5 kilobases. We hypothesize that these bands represent a double stranded RNA (dsRNA) corresponding to a virus with segmented genome, i.e. a reovirus. ds RNA was present in both males and females of the Samarkand;Raleigh$_{virus}$;Samarkand line and absent in Samarkand flies of both sexes as well as in male and female hybrids of reciprocal crosses between Samarkand and Samarkand;Raleigh$_{virus}$;Samarkand flies. To explain these results, we suggest that replication of cytoplasmic dsRNA fraction transmitted to hybrids by Samarkand;Raleigh$_{virus}$;Samarkand parents is suppressed by the second Samarkand chromosome, while Raleigh$_{virus}$ chromosome may carry a permissive allele of a gene (genes) allowing dsRNA production. No morphological characters different from wild type were noted in Samarkand;Raleigh$_{virus}$;Samarkand flies. Protocol of dsRNA extraction and purification was proposed. Cloning of putative virus genome for further sequencing and *in situ* hybridization; detection and analysis of virus particles and other experiments aimed to identify and describe the virus are in progress. After virus identification, we consider virus-host genome interactions to be of major interest for further investigation.

Short Communication C - Baculoviruses have long been used as natural agents to control insect pests in agriculture and forestry due to their notable safety for the environment and public health. Many baculoviruses, specifically the nucleopolyhedroviruses (NPV), have a very narrow host-range, killing one or two insect species. For example, the *Spodoptera exigua* multicapsid NPV (SeMNPV) and *Thysanoplusia orichalcea* multicapsid NPV (ThorMNPV) can kill only one or a few hosts and replicate well in a few insect cell lines (Cheng et al., 2005; Simon et al., 2004; Wang et al., 2008). A noteworthy exception to this generalization is the *Autographa californica* multicapsid NPV (AcMNPV) that has a wide host-range (>33 species) and can replicate well in many insect cell lines (Groner, 1986; McIntosh and Grasela, 1994). AcMNPV is the type species of the family *Baculoviridae*, and is also the first baculovirus whose genome was sequenced (Ayres et al., 1994). AcMNPV has a double-stranded circular DNA genome with a size of 134 kb (Ayres *et al.*, 1994). ThorMNPV and AcMNPV belong to the same group (I) and share about 80% DNA sequence homology (Cheng et al., 2005). SeMNPV belongs to the group II that shares less genome sequence homology to the group I viruses.

In: Insect Viruses: Detection, Characterization and Roles
Editors: Ch. J. Connell and D. P. Ralston
ISBN: 978-1-60692-965-0
© 2009 Nova Science Publishers, Inc.

Chapter 1

Insect Virus Proteins Involved in the Peroral Infectivity of the Viruses and Their Potential Practical Application in Pest Control

Wataru Mitsuhashi

National Institute of Agrobiological Sciences, Tsukuba, Ibaraki 305-8634, Japan

Abstract

The midgut is the initial place of infection of host insects by insect viruses that are perorally acquired. This initial (primary) infection leads to their replication in the main target tissues for replication in parts of the insects. Many studies have identified the range of proteins in many insect viruses, which are involved in their initial infection of the midgut. Some proteins are essential for peroral infection, and others only enhance the infection; some of the proteins initiate or enhance the infections by working the initial-infection tissue, called the midgut epithelium, and the others facilitate passage of the viruses across an acellular membrane, the peritrophic membrane (which is a barrier against the initial infection) toward the midgut epithelium by disrupting the membrane. Many studies have suggested that this enhancement of infectivity by some of the proteins is potentially applicable to pest control in agriculture and forestry, because the proteins are potential synergists for insect viral insecticides, and the co-administration of the proteins will drastically increase the efficacy of the viral pesticides. This will encourage far wider use of viral pesticides that have an advantage of being safer for vertebrates, plants, and the environment than synthetic pesticides. This chapter reviews basic and applied research on insect virus proteins associated with peroral infectivity, especially with respect to entomopoxviruses and granuloviruses, which belong to the families *Poxviridae* and *Baculoviridae*, respectively.

1. Introduction

Insects are the most abundant group of animals, with about 900 million insect species already discovered around the world. However, insect virus species have been found in very few insect species because of a shortage of studies designed to detect these viruses. Currently, known viruses that are found exclusively in insects range 12 families and several unclassified groups (*Iflavirus* etc.).

Insect viruses infect hosts perorally, and the midgut is the initial site of the infection. This initial (primary) infection leads to the replication of the viruses in the main target tissues for replication in other (sometimes the same) parts of the insects. Many studies have revealed that various proteins of insect viruses initiate or enhance their initial infection. Some proteins are essential for peroral infection, and others are not but enhance it; some of the proteins work the initial-infection tissue, called the midgut epithelium, and the others disrupt an acellular membrane, called the peritrophic membrane (PM), which is a physical barrier against initial infection, and thereby facilitate passage of the viruses across the PM toward the midgut epithelium.

Many studies have suggested that this enhancement of infectivity by some of the proteins is potentially applicable to pest control in agriculture and forestry, because the proteins are potential synergists of insect viral insecticides. Proteins that enhance or initiate infection have thus attracted the interest of insect virus researchers. The merit of viral insecticides is that they are far safer for vertebrates, plants and the environment than synthetic pesticides. The use, however, has been extremely limited to date, owing both to their high production costs and the narrow spectrum of pest insects that they can kill in comparison to synthetic insecticides. However, additives would decrease the amount of the viral insecticides used per unit area and thus reduce the cost of pest control. This will encourage far wider use of viral pesticides.

In this chapter, I review studies of the insect virus proteins associated with peroral infectivity, especially with respect to basic and applied research on the protein fusolin of entomopoxviruses (EPVs) and the protein enhancin of granuloviruses (GVs).

2. Spindles (Fusolin)

2.1. Properties of EPVs

Spindles are one type of inclusion body (i.e., a proteinaceous paracrystalline structure) whose major constituent protein, fusolin, is coded by an EPV gene (Figure 1A).

EPVs belong to the *Poxviridae*, which contains two subfamilies; the *Chordopoxvirinae*, which infect vertebrates, and the *Entomopoxvirinae*, which infect insects. The *Entomopoxvirinae* are subdivided into the following three genera: *Alphaentomopoxvirus* of Coleoptera, *Betaentomopoxvirus* of Lepidoptera and Orthoptera, and *Gammaentomopoxvirus* of Diptera. EPV virions are ovoid or brick-shaped (150-470 nm × 165-300 nm), and contain a large, linear, double-stranded DNA (225-380 kbp). EPVs replicate mainly in the fat bodies of host insects, and form another type of inclusion body, called spheroid, in the cytoplasm of host cells (Figure 1B). The spindles have not been found in hosts infected with *Gammaentomopoxvirus* species, or in insects infected with some *Betaentomopoxvirus* species. Spheroids and spindles vary in size; major axis is 5 to 20 μm in spheroids and 1 to 15 μm in spindles [King et al., 1998]. The major constitutive protein of the spheroid is spheroidin. The only protein reported other than fusolin in spindles is the endoplasmic reticulum-specific chaperone BiP [Lai-Fook & Dall., 2000]. The spheroids contain virions (Figure 1C), whereas the spindles do not. The spheroids may play a role in protection of the virions against ultraviolet radiation from the Sun and high temperature [Arif, 1995], and the function of the spindles may be to enhance EPV infectivity [Mitsuhashi et al., 2000, 2007; Mitsuhashi, 2002].

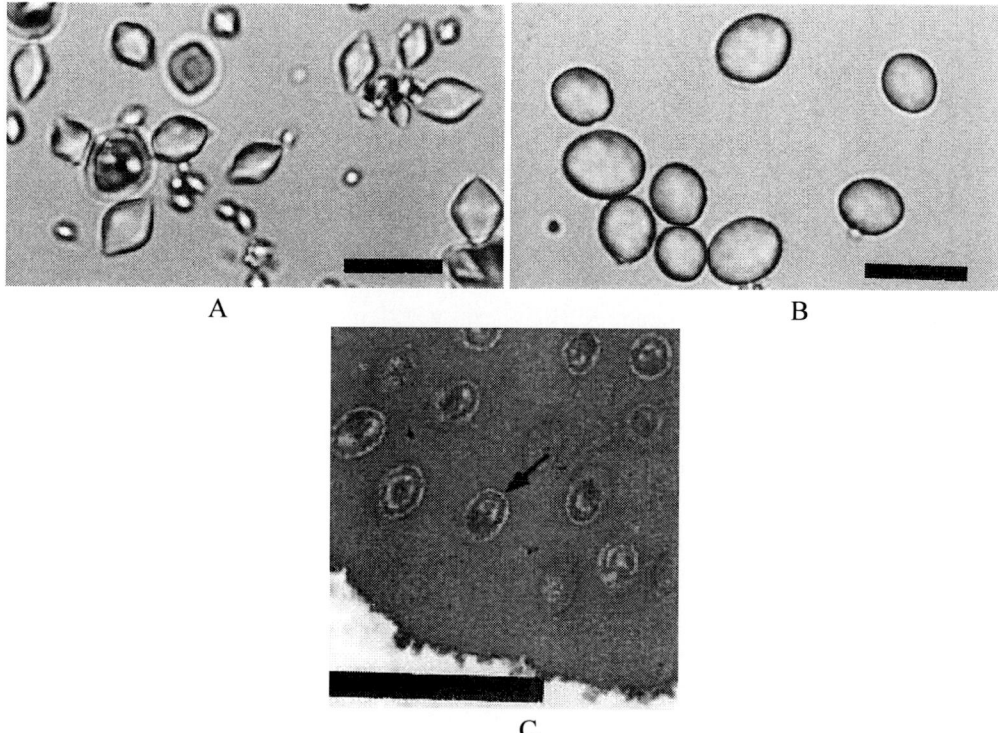

Figure 1. *Anomala cuprea* entomopoxvirus inclusion bodies: A) spindles; B) spheroids; C) sectional view of the spheroid. Bars in A and B indicate 10 μm and that in C 1μm. Arrow indicates a virion. Some virions show the characteristic of genus *Alphaentomopoxvirus*, in which the core is uniconcave.

2.2. Enhancement of Virus Infection

Xu & Hukuhara [1992] reported that the infectivity of *Pseudaletia unipuncta* nucleopolyhedrovirus (PuNPV) in *P. separata* larvae (Lepidoptera) was strongly enhanced by peroral administration of the viruses plus the spheroids from *Pseudaletia separata* EPV (PsEPV) or the solution of the spheroid matrix. NPVs belong to the family *Baculoviridae*. The factor in PsEPV spheroids that enhances the infectivity of PuNPV was reported to be a 38-kDa glycoprotein, which was called enhancing factor (EF) [Xu & Hukuhara, 1994].

Hukuhara et al. [1995] reported that the EF is present both in the PsEPV spindles and in the virions within the PsEPV spheroids, based on an immunoelectron microscopic analysis. The gene that encodes the EF was cloned from PsEPV and exhibited a significant similarity to the fusolin genes in other EPV species [Hayakawa et al., 1996], indicating that the EF is identical to PsEPV fusolin. A recombinant form of *Heliothis armigera* EPV (HaEPV) whose fusolin-coding sequence had been replaced with that of the green fluorescent marker protein, replicated in a cell culture, so the fusolin gene is not essential for virus replication [Olszewski & Dall, 2002]. Fusolin may not be essential for peroral infectivity, since several EPVs have been reported to lack in fusolin gene [Afonso et al., 1999; Bawden et al., 2000]. Fusolins vary from about 350 to 390 amino acid residues, and the N-terminus (about 15-20 aa residues) is a signal peptide that is cleaved from the protein to produce the mature form [Mitsuhashi et al., 1997]. This cleavage suggests that fusolin moves to the endoplasmic reticulum and crystallizes there. This suggestion is supported by the immunoelectron microscopic observation of HaEPV-infected host cells [Lai-Fook & Dall, 2000]. Some fusolins have one potential N-glycosylation site, and are glycoproteins [Xu & Hukuhara, 1994; Mitsuhashi et al., 1997]; fusolin from HaEPV is not glycosylated, despite the presence of a potential N-glycosylation site [Dall et al., 1993]. Mitsuhashi et al. [1998] showed that spindles of the EPV from the scarabaeid beetle *Anomala cuprea* (AcEPV), but not the spheroids, strongly enhanced *Bombyx mori* NPV (BmNPV) infectivity in *B. mori* larvae (Figure 2). Wijonarko & Hukuhara [1998] also showed that not only the spheroids of PsEPV, but also its spindles and virions, enhanced the infectivity of PuNPV in *P. separata* larvae. Furthermore several reports have shown enhanced infectivity of other NPV species by AcEPV or HaEPV spindles [Mitsuhashi & Sato, 2000; Chakraborty et al., 2004]. Based on bioassays using AcEPV and its spindles, EPV spindles have been demonstrated to enhance its infectivity, and facilitating EPV infection is thus thought to be the natural biological function of the spindles [Mitsuhashi et al., 2000; Mitsuhashi, 2002; Mitsuhashi et al., 2007].

EPV spindles may generally enhance NPV infectivity more than the occlusion bodies, called granules or capsules, of GV in the *Baculoviridae*: spindles of PsEPV, AcEPV, and HaEPV enhanced NPV infectivity by 10^6, 10^6 and 10^3 times, respectively [Mitsuhashi & Miyamoto, 2003; Wijonarko & Hukuhara, 1998; Chakraborty et al., 2004]. In contrast, GV granules containing enhancin (which is discussed in more detail in section 3 of this chapter) have been reported to enhance NPV infectivity by between 10^1 and 10^3 times, although PuNPV infection was enhanced by more than 10^4 times by *Pseudaletia unipuncta* GV (PuGV) granules [Tanada & Hukuhara, 1971; Goto, 1990; Hukuhara et al., 1987]. In addition, EPV spindles may have a wide activity spectrum of NPVs. For instance, AcEPV spindles strongly enhanced the infectivity of *Spilosoma imparilis* NPV (SiNPV) and BmNPV [Mitsuhashi et al., 1998; Mitsuhashi & Sato, 2000], even though these NPVs are not closely related taxonomically and AcEPV does not infect the hosts of either NPVs. The spectrum of activity of GV granules containing enhancin is not wide; they enhance the infectivity of NPVs that share hosts with the GVs and closely related NPVs to those NPVs. [Hukuhara et al., 1987; Derksen & Granados, 1988; Goto, 1990; Wang et al., 1994].

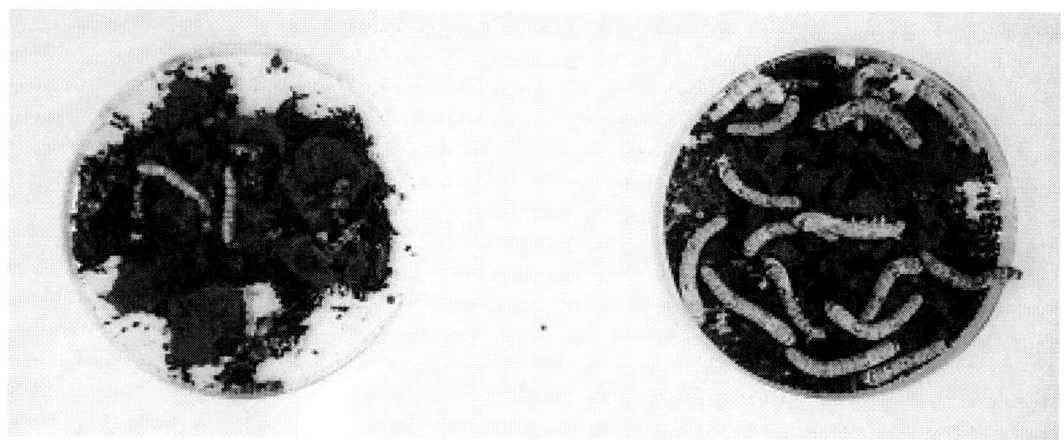

Figure 2. *Bombyx mori* larvae at five days after the termination of the administration of *Bombyx mori* nucleopolyhedrovirus (BmNPV) polyhedra containing the viruses with or without *Anomala cuprea* entomopoxvirus (AcEPV) spindles. Twenty third-instar larvae were group-administered BmNPV polyhedra with or without AcEPV spindles in an artificial diet in a petridish, and consumed the diet completely. The average numbers of BmNPV polyhedra and AcEPV spindles consumed by each larva are as follows. Left in the figure: polyhedra, 10; spindles, 2.5×10^6. Right: polyhedra, 1×10^5. A total of 20 larvae from each group were then transferred into a dish containing fresh artificial diet. All of the larvae fed with spindles died, and most darkened and were thus difficult to recognize in the image. In while, most larvae fed only polyhedra remained alive and clearly recognizable.

NPVs have two infectious phenotypes: an occluded virus (polyhedron-derived virus or occlusion-derived virus) (ODV) and a non-occluded virus (budded virus) (NOV or BV). ODVs are occluded by polyhedra and NOVs are not. NOVs are rarely infectious through the insect midgut as a result of peroral administration. The above-mentioned studies that revealed the enhancement of NPV infection used polyhedra that contained viruses. However, Furuta et al. [2001] showed that NOVs of BmNPV (both of wild type NPV and polyhedrin-negative recombinant NPV) become highly infectious to *B. mori* larvae even through peroral inoculation when they are administered with AcEPV spindles. This finding suggests that polyhedrin-negative recombinant baculovirus will be a strong tool for large-scale inoculation experiments and for mass-production biofactories using the recombinant viruses. This is because it is easier to perorally inoculate these recombinant viruses into a large number of hosts than to use the conventional method of injecting the viruses into host haemocoel one by one.

The enhancing-activity of NPV infectivity by AcEPV spindles are highly stable against various abiotic factors such as high temperature, ultraviolet radiation, formaldehyde, ethyl alcohol, and the bactericide benzalkonium chloride [Mitsuhashi et al., 2008]. For example, AcEPV spindles retain high activity even after heating at 95 °C for 30 min, whereas PuGV granules were inactivated by heating at 85 °C for 10 min [Tanada, 1959]. These results indicate that the spindles are relatively stable, which is necessary for their use as a synergist of viral pesticides.

A few NPVs form spindle-shaped inclusion bodies [Huger & Krieg, 1968; Adams & Wilcox, 1968; Li et al., 2000; Chakraborty et al., 2005], and the constituent protein is generally called GP37 [Gross et al., 1993]. GP37 is also found in baculoviruses that do not form the spindle-shaped bodies, and shows significant similarity to the fusolin of EPVs

[Gross et al., 1993; Dall et al., 1993]. Cheng et al. [2001] reported that GP37 is not essential for *Autographa californica* NPV (AcNPV) replication and did not enhance the infectivity of the NPV, based on the assays using a mutant with inactivated *gp37* of AcNPV and *gp37*-deletion mutant [Cheng et al., 2001]. The spindle-shaped bodies (diamond-shaped bodies) of *Galleria mellonella* NPV were reported to enhance the peroral infectivity of *Mythimna separata* NPV species, suggesting a biological function similar to that of the EPV spindles [Chakraborty et al., 2005]; further study will determine the function of the spindle-shaped bodies of the NPVs.

2.3. Mode of Action of Infectivity-Enhancement

The initial process of infection of host insects by DNA viruses that are occluded by occlusion bodies begins with ingestion of the occlusion bodies (i.e., polyhedra, granules, and spheroids), which dissolve in the alkaline midgut juices of the insects. The liberated virions by this process then pass through the PM and into the ectoperitrophic space. The virions then attach to the microvilli of the midgut columnar cells, and the viral envelopes fuse to them. Finally, the nucleocapsids of baculoviruses or the cores with lateral bodies of EPVs inside the viral envelopes enter the host cells. In this infection process, the PM was shown to act as a barrier between the NPV virions and the microvilli in some lepidopteran insects [Wang & Granados, 1997, 2000, 2001; Peng et al., 1999], and is considered to likely play a similar role in other insects. The PM is an acellular membrane that lines the midgut lumen in the form of a tube, extending from the anterior midgut to the hindgut [Derksen & Granados, 1988]. The PM is primarily composed of chitin and proteins, including glycoproteins and proteoglycans [Wang & Granados, 2001].

Mitsuhashi & Miyamoto [2003] showed that the PM of *B. mori* larvae was disintegrated by AcEPV spindles, and the PM conformation of *B. mori* not treated with the spindles appeared to physically prevent passage of BmNPV to a considerable degree [Mitsuhashi & Murakami, 2007]. Thus the disintegration of the PM is thought to be one mode of action by which EPV spindles enhance NPV infectivity. Furthermore, the PM of *A. cuprea* larvae was found to be disrupted after the larvae were fed AcEPV spindles, thereby facilitating infection of the host's midgut by the EPV [Mitsuhashi et al., 2007]. This strongly suggests that the mode of action of enhancement of viral infection is similar in NPV and EPV infections. Hukuhara & Wijonarko [2001] and Hukuhara et al. [2001] showed that the EF (PsEPV fusolin) enhanced the fusion between PuNPV virions and cultured insect cells derived from haemocytes, and speculated that the enhancement of NPV infection in *P. separata* larvae was resulted from enhanced fusion of the virions to the microvilli of midgut columnar cells mediated by the EF. Thus, this enhanced fusion may be another mode of action of the spindles (i.e. of fusolin) in the enhancement of NPV infection; further studies are required to investigate whether this phenomenon occurs in vivo.

Fusolins show a distant similarity to some bacterial chitin-binding proteins, including Cbp1 from *Alteromonas* sp. strain O-7 and CHB1 from *Streptomyces olivaceoviridis*, to CBP from *S. halstedii*, and to ChiBs from some bacteria, including *Pseudoalteromonas* sp. strain S9 and *Salinivibrio costicola* [Li et al., 2003; Tsujibo et al., 2002; Vaaje-Kolstad et al.,

2005]. Fusolins and GP37s and have some conserved regions that constitute a chitin-binding domain, like the corresponding regions of the proteins of these bacteria [Li et al., 2003; Tsujibo et al., 2002; Vaaje-Kolstad et al., 2005], and the domain of fusolins, GP37s, and some of these bacterial proteins have been specifically classified into chitin-binding domain 3 (InterPro database). Indeed, *Spodoptera litura* NPV GP37 showed chitin-binding ability [Li et al., 2003], and AcEPV fusolin has been demonstrated to bind to chitin by in an in vitro assay [Takemoto et al., 2008]. Therefore, fusolin may inhibit new formation of the PM by binding to newly synthesized chitin and thus inhibiting the binding of PM proteins to the chitin. Alternatively, it may bind to chitin in the PM and alter the PM conformation, leading to the disruption of the membrane.

3. Enhancin (Granules)

3.1. Properties of GVs

The family *Baculoviridae* is composed of two genera: *Nucleopolyhedrovirus* and *Granulovirus*. Viruses in the former genus and the latter are called NPVs and GVs, respectively, and viruses in this family are baculoviruses. Baculovirus virions are large and rod-shaped, and contain a circular, double-stranded DNA (80-180 kbp). Baculoviruses only infect arthropods. In general, they multiply in almost all types of organs and tissues of host insects (Figure 3).

Each GV ODV is occluded within an ovicylindrical occlusion body (granule) composed primarily of a protein called granulin. Granules measure approximately 0.13 × 0.50 μm in size. Each virion typically contains a single nucleocapsid within a single envelope, and the ODV mature in the nucleus of infected cells. Nucleocapsids are 30-60 × 250-300 nm and contain a DNA approximately 80 to 180 kbp. Species in this genus have been isolated only from Lepidoptera [Federici, 1997].

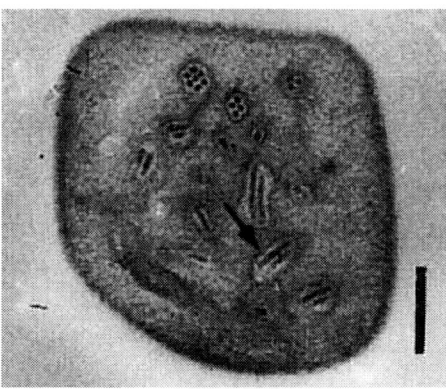

Figure 3. Section of a nucleopolyhedrovirus polyhedron from *Cyclophragma yamadai* (Lepidoptera: Lasiocampidae). Arrow indicates a virion. Bar indicates 0.5 μm. The sections of rod-shaped virions vary in shape, depending on the axis of the section.

Enhancin has been identified in many GVs and some NPVs [Lapointe et al., 2004] and is localized in the granule matrix [Tanada, 1985]. However, *Lymantria dispar* NPV (LdNPV) enhancins were shown to be present in ODV envelopes in association with nucleocapsids [Slavicek & Popham, 2005].

3.2. Enhancement of Baculovirus Infectivity

The first study concerning the enhancement of baculovirus infectivity by enhancin was by Tanada [1959], who found enhancement of PuNPV infection in *P. unipuncta* larvae after feeding them PuGV granules. The factor in PuGV responsible for enhanced infectivity, which was designated synergistic factor (SyF), was identified as a 126K enzymatic protein that had phospholipids as an enhancing component [Tanada, 1975; Yamamoto & Tanada, 1978a]. The SyF is virtually identical to enhancin (See section 3.3). Furthermore, other reports indicated that a minor factor in a fraction of purified granulin from *Trichoplusia ni* GV (TnGV) granules enhanced the infectivity of AcNPV in *T. ni* larvae [Derksen & Granados, 1988] and that *Xestia c-nigrum* GV granules enhanced *Xestia c-nigrum* NPV infection in *X. c-nigrum* larvae [Goto, 1990]. There have also been reports of infectivity enhancement by purified enhancins in TnGV enhancin–AcNPV and PuGV enhancin (SyF)–PuNPV combinations [Hara et al., 1976; Gallo et al., 1991]. Tanada [1975] also reported that infectivity of two strains of PuGV were enhanced by SyF, suggesting that natural biological function of SyF is to enhance the infectivity of GV.

In addition to enhancin, some proteins in granules such as PIF and "11K group proteins" mentioned below (See section 4 of this chapter), have been reported to be involved in peroral infectivity. Therefore, previously reported enhancement by granules may be due to the integration of action by proteins associated with oral infectivity, including enhancin.

LdNPV was the first NPV in which an enhancin gene was identified, and its genome contains two enhancin genes and, by bioassays using enhancin gene(s) deleted or inactivated recombinant LdNPV, it was demonstrated that both are not essential for LdNPV replication and contribute to the enhancement of peroral infectivity [Bischoff and Slavicek, 1997; Popham et al., 2001]. From these results together with those by Tanada [1975] that the infectivity of PuGV was enhanced by SyF, natural biological function of enhancin is thought to enhance the infectivity of GVs.

3.3. Mode of Action of Infectivity-Enhancement

Tanada et al. [1983] and Tanada [1985] suggested that the enhancement of PuNPV infectivity resulted from the enhanced fusion of the NPV virions to the microvilli of midgut columnar cells caused by the SyF, based on the biochemical studies of the protein and on electron microscopy of the microvilli. This suggestion was supported by Kozuma & Hukuhara [1994], who showed that the SyF of PuGV enhanced the fusion between *Trichoplusia ni* NPV virions and cultured insect cells.

On the other hand, Derksen & Granados [1988] found that the protein composition of the PM was altered and the PM became fragile, in *T. ni* larvae that were fed TnGV granules, showing a possibility that the infectivity-enhancement of AcNPV by the granules is due to the disruption of the PM. The gene encoding the protein factor of TnGV that enhanced viral infectivity was cloned and sequenced [Hashimoto et al., 1991]. This protein in TnGV was designated enhancin [Corsaro et al., 1993]. Comparison of a gene coding for SyF in PuGV and the TnGV enhancin gene revealed that the two proteins are virtually identical each other [Roelvink et al., 1995].

Wang et al. [1994] found no specific binding sites for TnGV enhancin on the microvilli of the midgut of *T. ni*, *Helicoverpa zea*, or *Spodoptera exigua*, although they found a binding site in *P. unipuncta*, suggesting that specific binding of enhancin to the midgut cell membrane is not necessary for enhancement of NPV infection in insects.

TnGV enhancin was identified as a metalloprotease that degraded the intestinal mucin of the PM [Lepore et al., 1996; Wang & Granados, 1997]. This degradation increases the permeability of the PM to NPVs, helping them to reach the microvilli and enhancing their infectivity [Wang & Granados, 1998; Peng et al., 1999].

4. Other Proteins

Other proteins of insect viruses are involved in initial NPV infectivity in the midgut. The deletion of the genes encoding these proteins from some NPVs resulted in the reduction or lost of the peroral infectivity of the viruses, and thus they are identified as proteins involved in peroral infectivity.

P74 is an ODV envelope protein [Faulkner et al., 1997] and facilitates the specific binding of ODV to midgut epithelia [Hass-Stapleton et al., 2004]. This protein is essential for peroral infection [Kuzio et al., 1989; Yao et al., 2004]. The C-terminus of this protein is anchored in the virus envelope [Slack et al., 2001] and the N-terminus is exposed on the virion surface [Faulkner et al., 1997]. P74 is cleaved on its N-terminus by midgut trypsins [Slack & Lawrence, 200]. The central transmembrane domain region may be involved in membrane fusion and may be exposed by protease cleavage [Slack & Arif, 2007].

PIF-1 is an ODV structural protein [Kikhno et al., 2002]. PIF-1 and PIF-2 are 60-kDa and 44-kDa proteins, respectively, and contain prominent N-terminal transmembrane motifs that may work as transmembrane anchors [Slack & Arif, 2007]. Both are involved in the binding of ODV to midgut cells [Ohkawa et al., 2005]. PIF-3, a 23-kDa protein, is also an ODV structural protein encoded by AcMNPV ORF *Ac115* and does not mediate specific binding to the epithelia [Ohkawa et al., 2005]. These three proteins are essential for peroral infection. [Kikhno et al., 2002; Pijlman et al., 2003; Ohkawa et al., 2005].

There are proteins that promote rapid transmission of BV that are produced in the midgut cells to tracheolar cells and thereby enhance the peroral infectivity [Zhang et al., 2005]. These factors are diverse and includes PE38 [Milks et al., 2003] and GP64 [Washburn et al., 2003; Zhang et al., 2004].

Ac145 and Ac150 of AcNPV belonging to a group of the 11K proteins are considered to be ODV envelope proteins that are not essential for peroral infection [Lapointe et al., 2004;

Zhang et al., 2005]. Ac145 and Ac150 were shown to enhance peroral infectivity of AcNPV in bioassays using deletion mutant viruses of these proteins [Lapointe et al., 2004; Zhang et al., 2005]. The 11K genes were found in many baculoviruses and HaEPV and are classified into two subtypes [Dall et al., 2001]; the effect of the 11K proteins in the EPV has not yet been examined. These gene products are predicted to harbor a hydrophobic N-terminal domain and a core C6 motif of conserved cysteine residues in a well-defined spacing pattern [Dall et al., 2001]. The C6 motifs are recognized to be chitin-binding motifs or peritrophin-A domains that are popular among mucins, peritrophins, and chitinases [Tellam et al., 1999]. However, enhancement of perorall infectivity by Ac150 protein did not involve the PM [Zhang et al., 2005].

5. Application of Viral Proteins in Pest Control

5.1. Viral Insecticides and Improvement of Their Infectivity

Unlike most synthetic pesticides, insect-viral pesticides are safe for vertebrates (including humans and livestock), plants, most beneficial insects, and the environment when they are used to control specific insect pests in agriculture and forestry [Tanada & Kaya, 1993]. Baculoviruses are considered to be especially safe as viral pesticides, because they infect only arthropods. Baculoviruses thus have attracted much attention as potential biological control agents. However, the use of the viral pesticides has been very limited, compared with synthetic pesticides. The reasons for this are as follows. (1) The commercial production costs of viral pesticides are much higher than those of synthetic pesticides, because their production is much more labor-intensive. (2) The period from infection to death is relatively long, and during that time, the pests continue to consume the crops. (3) The spectrum of pests killed by each virus species is narrow, so it may be necessary to develop a different insecticide for each insect or group of insects. (4) Viruses are easily inactivated by abiotic factors such as ultraviolet radiation from the Sun and high temperature. (5) Application of viral pesticides to pests must be carefully timed to provide the optimal effect.

To promote the wider use of viral pesticides, it is important to overcome at least some of these drawbacks. One major solution would be a dramatic strengthening of the ability of viruses to kill pests by improving their infectivity, speed of killing, and host range. Until now, two major methods of strengthening the ability to kill pests have been studied. One is the use of recombinant baculoviruses constructed by introducing foreign genes into the virus genome to increase the speed of killing and to expand the host range. However, no recombinant baculoviruses have been registered as viral pesticides, owing to a lack of public acceptance of the use of such genetically modified organisms in fields. Another method is the use of synergistic co-agents that can enhance the peroral infectivity of viruses and somewhat accelerates their speed of killing, thereby reducing the amount of viruses needed per unit area. Herein, the use of viral synergists would lower the costs of viral pest control and therefore would promote the use of the viral pesticides. Materials that have been reported to show a synergistic effect on the *per os* infectivity of insect viruses by oral administration of the materials are listed in Table 1. Among them, stilbene-derived optical brighteners, enhancin, and fusolin have been comparatively well studied. In general, natural synergists are likely to be less harmful to vertebrates and the environment than synthetic compounds.

Table 1. Materials that have been shown to increase the peroral infectivity of insect viruses after peroral administration to host larvae

Material (scientific or commercial name)	Reference
Phosphatidyl choline (lecithin)	Yamamoto & Tanada [1978b], Tuan & Hou [1988]
Dodecylamine hydrochloride	Yamamoto & Tanada [1978b]
Cetyltrimethylammonium bromide	Yamamoto & Tanada [1978b]
Boric acid	Shapiro & Bell [1982]
Stilbene-derived optical brightner	
Leucophor BS	Shapiro & Robertson [1992]
Leucophor BSB	Shapiro & Robertson [1992]
Phorwite AR	Shapiro & Robertson [1992]
Phorwite RKH	Shapiro & Robertson [1992]
Tinopal LPW (Fluorescent brightener 28, Calcofluor white M2R)	Shapiro & Robertson [1992]
(Tinopal UNPA-GX)*	Arakawa et al. [2000]
Tinopal UNPA-GX derivatives (SN, S, 3BS, FB, B, CR, PAS, LSK)	Okuno et al. [2003]
Blankophor P167	Shapiro & Hamm [1999]
Blankophor BBH	Webb et al. [1996]
Chlorfluazuron	Guo et al. [2007]
Flufenoxuron	Arakawa [2002]
Nikkomycin Z	Arakawa & Sugiyama [2002]
Orthocide (captan)	Arakawa [2003]
Polyoxin complex	Arakawa [2003]
Chitinase	Shapiro et al. [1987]
Granulovirus (GV) granules (enhancin)	
Pseudaletia unipuncta GV granules	Tanada [1959]
Trichoplusia ni GV granules	Derksen & Granados [1988]
Xestia c-nigrum GV granules	Goto [1990]
Entomopoxvirus (EPV) inclusion bodies (fusolin)	
Pseudaletia separata EPV virions, spheroids and spindles	Xu & Hukuhara [1992], Wijonarko & Hukuhara [1998]
Anomala cuprea EPV spindles	Mitsuhashi et al. [1998]
Helicoverpa armigera EPV (*Heliothis armigera* EPV) spindles	Chakraborty et al. [2004]

*Tinopal UNPA-GX has almost the same molecular structure as Tinopal LPW, and can thus be regarded as essentially the same material.

Some viral proteins, including those mentioned in this chapter other than enhancin and fusolin, might also be used as the synergists if they can be produced abundantly in vitro or by other organisms. Rao et al. [2004] showed that the *B. mori* PM was disrupted in vitro by AcMNPV ChiA, a chitinase belonging to the ChiA class of bacterial chitinases, that had been expressed in *Escherichia coli,* although they did not examine whether it enhanced NPV infectivity; however, the membrane disruption suggest that ChiA is a potential synergist for viral insecticides. Although the ChiA deletion mutant AcMNPV did not significantly change the peroral infectivity in *T. ni* larvae compared with wild-type NPV [Hawtin et al., 1997], it may have weakly enhanced infectivity.

The feeding stimulant Coax (AgroSolutions, San Marcos, CA, USA) is composed of vegetable flours, oils, and sugars, and has shown some potential for improving the rate of killing of some pests [Farrar et al., 1999]. When NPV polyhedra mixed with Coax were sprayed on crops, insect pests were stimulated to feed, thereby increasing the number of polyhedra they consumed compared with that of polyhedra insects consumed by feeding of crops on which only polyhedra were sprayed, thereby increasing mortality rates. Therefore the material does not appear to enhance virus infectivity *per se*. A distinction should also be made between the ability of some materials to protect viral infectivity and the ability of materials to enhance infectivity. For example, lignin is a protective material [McGuire et al., 2001], and is used in the formulation of viral pesticides to maintain their infectivity. Stilbene-derived optical brighteners play both roles: they enhance NPV infection but also maintain infectivity by protecting the viruses from ultraviolet light [Shapiro & Robertson, 1992].

5.2. Developing Efficient Mass Production of the Proteins

Until now, only spindles (fusolin) and granules (enhancin) have been produced in significant quantities in the hosts of viruses for enhancing virus infectivity.

For mass production of these proteins, it is not always practical to use host insects. Such production is very labor-intensive, because the approach requires mass-rearing of insects followed by protein purification, and this increases production costs. AcEPV spindle (fusolin) production using *A. cuprea* larvae especially poses serious problems. It is difficult to group-rear the larvae, the method that is usually used to rear lepidopteran larvae, because this species has a strong cannibalistic tendency. This means that more laborious individual rearing must be used [Mitsuhashi, 2006]. In addition, it takes at least one month for the formation of many spindles in the larvae after inoculation with the viruses. Thus, an alternative method, such as the use of other organisms that require less labor and time to rear or culture, or in vitro production, is needed for the practical production of the proteins. The expression of the protein genes in crop plants after introduction of these genes into the crops is another potential method, since it is unnecessary to prepare the proteins as additives of viral pesticides.

Hukuhara et al. [1999] introduced a fusolin gene isolated from PsEPV into rice cultivars and observed an approximately 360-fold enhancement of PuNPV infectivity in *P. separata* larvae that were fed on transgenic plants that expressed the fusolin gene. In addition, Hukuhara et al. [2001] expressed fusolin gene in *E. coli* that had been transformed by PsEPV fusolin gene, and reported that the protein produced by this gene enhanced the infectivity of PuNPV in *P. separata* larvae by nearly 270 times. Our group has expressed the AcEPV fusolin gene in a baculovirus expression system and found that the resulting protein enhanced BmNPV infection of *B. mori* larvae by about 300 times [Takemoto et al., 2008]. The degrees of infectivity enhancement by the alternative methods were generally lower than those achieved by using spindles. From the practical perspective, it is important to elucidate the cause of this phenomenon and to find a solution.

Lepore et al. [1996] showed that TnGV enhancin produced in a baculovirus expression system enhanced AcNPV infection in *Trichoplusia ni* larve. Hayakawa et al. [2000] introduced TnGV enhancin gene into a strain of tobacco (*Nicotiana tabacum*) and expressed the gene in it. The infection of *S. exigua* larvae by AcNPV was enhanced by 10 times compared with that in the control when the larvae were fed the lyophilized leaves; thus the effect of enhancement by using a transgenic plant was weak.

6. Conclusion

EPVs and baculoviruses have gained different tools (proteins) for the common purpose of enhancing their initial infection of the host midgut, as described above, i.e., fusolin (spindles) and enhancin targeting the PM, although the related proteins, such as the 11K proteins, have also been obtained. Thus, insect viruses in other taxa likely have evolved other or similar gene products for their oral infectivity.

To date, these viral proteins have not been well characterized, although enhancin has been comparatively well studied, and fusolin is being studied. Details of mode of action, of location in occlusion bodies and of characteristics of the structures, and potential other modes of action, remain to be studied. Studies are required to discover other proteins of EPVs, which are involved in peroral infectivity.

As noted in this review, some of the proteins show potential as useful synergists for viral insecticides, and thus a number of studies are expected for putting them to practical use. One of the most important themes in such research should be to establish easier, lower-cost methods of mass production of the proteins.

References

Adams, J. R., & Wilcox, T. A. (1968). Histopathology of the almond moth, *Cadra cautella*, infected with a nuclear-polyhedrosis virus. *Journal of Invertebrate Pathology, 12*, 269-274.

Afonso, C. L., Tulman, E. R., Lu, Z., Oma, E., Kutish, G. F., & Rock, D. L. (1999). The genome of *Melanoplus sanguinipes* entomopoxvirus. *Journal of Virology, 73*, 533-552.

Arakawa, T. (2002). Promotion of nucleopolyhedrovirus infection in larvae of the silkworm, *Bombyx mori* (Lepidoptera: Bombycidae) by flufenoxuron. *Applied Entomology and Zoology, 37*, 7-11.

Arakawa, T. (2003). Chitin synthesis inhibiting antifungal agents promote nucleopolyhedrovirus infection in silkworm, *Bombyx mori* (Lepidoptera: Bombycidae) larvae. *Journal of Invertebrate Pathology, 83,* 261-263.

Arakawa, T., Kamimura, M., Furuta, Y., Miyazawa, M., & Kato, M. (2000). Peroral infection of nuclear polyhedrosis virus budded particles in the host, *Bombyx mori* L., enabled by an optical brightener, Tinopal UNPA-GX. *Journal of Virological Methods, 88,* 145-152.

Arakawa, T., & Sugiyama, M. (2002). Promotion of nucleopolyhedrovirus infection in larvae of the silkworm, *Bombyx mori* (Lepidoptera: Bombycidae) by an antibiotic, nikkomycin Z. *Applied Entomology and Zoology, 37,* 393-397.

Arif, B. M. (1995). Recent advances in the molecular biology of entomopoxviruses. *Journal of General Virology, 76,* 1-13.

Bawden, A. L., Glassberg, K. J., Diggans, J., Shaw, R., Farmerie, W., & Moyer, R. W. (2000). Complete genomic sequence of the *Amsacta moorei* entomopoxvirus: analysis and comparison with other poxviruses. *Virology, 274,* 120-139.

Bischoff, D. S., & Slavicek, J. M. (1997). Molecular analysis of an enhancin gene in the *Lymantria dispar* nuclear polyhedrosis virus. *Journal of Virology, 71,* 8133-8140.

Chakraborty, M., Narayanan, K., & Sivaprakash, M. K. (2004). *In vivo* enhancement of nucleopolyhedrovirus of oriental armyworm, *Mythimna separata* using spindles from *Helicoverpa armigera* entomopoxvirus. *Indian Journal of Experimental Biology, 42,* 121-123.

Chakraborty, M., Narayanan, K., Suryanarayana, V. V. S., & Sivaprakash, M. K. (2005). Enhancement of nucleopolyhedrovirus of oriental armyworm, *Mythimna separata* (Lepidoptera: Noctuidae) using diamond shaped inclusion bodies of *Galleria mellonella* NPV (Lepidoptera: Pyralidae). *Entomon, 30,* 343-346.

Cheng, X.-W., Krell, P. J., & Arif, B. M. (2001). P34.8 (GP37) is not essential for baculovirus replication. *Journal of General Virology, 82,* 299-305.

Corsaro, B. G., Gijzen, M. R., Wang, P., & Granados, R. R. (1993). Baculovirus enhancing proteins as determinants of viral pathogenesis. In N. E. Beckage, S. N. Thomson, & B. A. Federici, (Eds.), *Parasites and Pathogens of Insects vol. 2 Pathogens* (pp.127-145). San Diego: Academic Press.

Dall, D., Luque, T., & O'Reilly, D. (2001). Insect-virus relationships: sifting by informatics. *BioEssays, 23,* 184-193.

Dall, D., Sriskantha, A., Vera, A., Lai-Fook, J., & Symonds, T. (1993). A gene encoding a highly expressed spindle body protein of *Heliothis armigera* entomopoxvirus. *Journal of General Virology, 74,* 1811-1818.

Derksen, A. C. G., & Granados, R. R. (1988). Alteration of a lepidopteran peritrophic membrane by baculoviruses and enhancement of viral infectivity. *Virology, 167,* 242-250.

Farrar, R. R. Jr., Ridgway, R. L., & Dively, G. P. (1999). Activity and persistence of the nuclear polyhedrosis virus of the celery looper (Lepidoptera: Noctuidae) with a feeding stimulant and a stilbene-derived enhancer. *Journal of Entomological Science, 34,* 369-380.

Faulkner, P., Kuzio, J., Williams, G. V., & Wilson, J. A. (1997). Analysis of p74, a PDV envelope protein of *Autographa californica* nucleopolyhedrovirus required for occlusion body infectivity in vivo. *Journal of General Virology, 78,* 3091-3100.

Federici, B. A. (1997). Baculovirus pathogenesis. In L. K. Miller (Ed.), *The Baculoviruses* (pp. 33-59). New York: Plenum Press.

Furuta, Y., Mitsuhashi, W., Kobayashi, J., Hayasaka, S., Imanishi, S., Chinzei, Y., & Sato, M. (2001). Peroral infectivity of non-occluded viruses of *Bombyx mori* nucleopolyhedrovirus and polyhedrin-negative recombinant baculoviruses to silkworm larvae is drastically enhanced when administered with *Anomala cuprea* entomopoxvirus spindles. *Journal of General Virology, 82,* 307-312.

Gallo, L. G., Corsaro, B. G., Hughes, P. R., & Granados, R. R. (1991). In vivo enhancement of baculovirus infection by the viral enhancing factor of a granulosis virus of the cabbage looper, *Trichoplusia ni* (Lepidoptera: Noctuidae). *Journal of Invertebrate Pathology, 58,* 203-210.

Goto, C. (1990). Enhancement of a nuclear polyhedrosis virus (NPV) infection by a granulosis virus (GV) isolated from the spotted cutworm, *Xestia c-nigrum* L. (Lepidoptera: Noctuidae). *Applied Entomology and Zoology, 25,* 135-137.

Gross, C. H., Wolgamot, G. M., Russell, R. L. Q., Pearson, M. N., & Rohmann, G. F. (1993). A 37-kilodalton glycoprotein from a baculovirus of *Orgyia pseudotsugata* is localized to cytoplasmic inclusion bodies. *Journal of Virology, 67,* 469-475.

Guo, H.-F., Fang, J.-C., Liu, B.-S., Wang, J.-P., Zhong, W.-F., & Wan, F.-H. (2007). Enhancement of the biological activity of nucleopolyhedrovirus through disruption of the peritrophic matrix of insect larvae by chlorfluazuron. *Pest Management Science, 63,* 68-74.

Hara, S., Tanada, Y., & Omi, E. M. (1976). Isolation and characterization of a synergistic enzyme from the capsule of a granulosis virus of the armyworm, *Pseudaletia unipuncta*. *Journal of Invertebrate Pathology, 27,* 115-124.

Hashimoto, Y., Corsaro, B. G., & Granados, R. R. (1991). Location and nucleotide sequence of the gene encoding the viral enhancing factor of the *Trichoplusia ni* granulosis virus. *Journal of General Virology, 72,* 2645-2651.

Hass-Stapleton, E. J., Washburn, J. O., & Volkman, L. E. (2004). P74 mediates specific binding of *Autographa californica* M nucleopolyhedrovirus occlusion-derived virus to primary cellular targets in the midgut epithelia of *Heliothis virescens* larvae. *Journal of Virology, 78,* 6786-6791.

Hawtin, R. E., Zarkowska, T., Arnold, K., Thomas, C. J., Gooday, G. W., King, L. A., Kuzio, J. A., & Possee, R. D. (1997). Liquefaction of *Autographa californica* nucleopolyhedrovirus-infected insects is dependent on the integrity of virus-encoded chitinase and cathepsin genes. *Virology, 238,* 243-253.

Hayakawa, T., Shimojo, E., Mori, M., Kaido, M., Furusawa, I., Miyata, S., Sano, Y., Matsumoto, T., Hashimoto, Y., & Granados, R. R. (2000). Enhancement of baculovirus infection in *Spodoptera exigua* (Lepidoptera: Noctuidae) larvae with *Autographa californica* nucleopolyhedrovirus or *Nicotiana tabacum* engineered with a granulovirus enhancin gene. *Applied Entomology and Zoology, 35,* 163-170.

Hayakawa, T., Xu, J., & Hukuhara, T. (1996). Cloning and sequencing of the gene for an enhancing factor from *Pseudaletia separata* entomopoxvirus. *Gene, 177*, 269-270.

Huger, A. M., & Krieg, A. (1968). On spindle-shaped cytoplasmic inclusions associated with a nuclear polyhedrosis of *Choristoneura murinana*. *Journal of Invertebrate Pathology, 12*, 461-462.

Hukuhara, T., Hayakawa, T., & Wijonarko, A. (1999). Increased baculovirus susceptibility of armyworm larvae feeding on transgenic rice plants expressing an entomopoxvirus gene. *Nature Biotechnology, 17*, 1122-1124.

Hukuhara, T., Hayakawa, T., & Wijonarko, A. (2001). A bacterially produced virus enhancing factor from an entomopoxvirus enhances nucleopolyhedrovirus infection in armyworm larvae. *Journal of Invertebrate Pathology, 78*, 25-30.

Hukuhara, T., Tamura, K., Zhu, Y., Abe, H., & Tanada, Y. (1987). Synergistic factor shows specificity in enhancing nuclear polyhedrosis virus infections. *Applied Entomology and Zoology, 22*, 235-236.

Hukuhara, T., & Wijonarko, A. (2001). Enhanced fusion of a nucleopolyhedrovirus with cultured cells by a virus enhancing factor from an entomopoxvirus. *Journal of Invertebrate Pathology, 77*, 62-67.

Hukuhara, T., Yano, K., Xu, J., Tomita, M., & Miyajima, S. (1995). Detection of a virus enhancing factor in inclusion bodies of an entomopoxvirus by immunoelectron microscopy. *Journal of Invertebrate Pathology, 65*, 315-317.

Kikhno, I., Gutiérrez, S., Croizier, L., Croizier, G., & Ferber, M. L. (2002). Characterization of *pif*, a gene required for the *per os* infectivity of *Spodoptera littoralis* nucleopolyhedrovirus. *Journal of General Virology, 83*, 3013-3022.

King, L. A., Wilkinson, N., Miller, D. P., & Marlow, S. A. (1998). Entomopoxviruses. In L. K. Miller, & L. A. Ball (Eds.), *The Insect Viruses* (pp. 1-29). New York and London: Plenum Press.

Kozuma, K., & Hukuhara, T. (1994). Fusion characteristics of a nuclear polyhedrosis virus in cultured cells: time course and effect of a synergistic factor and pH. *Journal of Invertebrate Pathology, 63*, 63-67.

Kuzio, J., Jaques, R., & Faulkner, P. (1989). Identification of p74, a gene essential for virulence of baculovirus occlusion bodies. *Virology, 173*, 759-763.

Lai-Fook, J., & Dall, D. J. (2000). Spindle bodies of *Heliothis armigera* entomopoxvirus develop in structures associated with host cell endoplasmic reticulum. *Journal of Invertebrate Pathology, 75*, 183-192.

Lapointe, R., Popham, H. J. R., Straschil, U., Goulding, D., O'Reilly, D. R., & Olszewski, J. A. (2004). Characterization of two *Autographa californica* nucleopolyhedrovirus proteins, Ac145 and Ac150, which affect oral infectivity in a host-dependent manner. *Journal of Virology, 78*, 6439-6448.

Lepore, L. S., Roelvink, P. R., & Granados, R. R. (1996). Enhancin, the granulosis virus protein that facilitates nucleopolyhedrovirus (NPV) infections, is a metalloprotease. *Journal of Invertebrate Pathology, 68*, 131-140.

Li, X., Barrett, J., Pang, A., Klose, R. J., Krell, P. J., & Arif, B. M. (2000). Characterization of an overexpressed spindle protein during a baculovirus infection. *Virology, 268*, 56-67.

Li, Z., Li, C., Yang, K., Wang, L., Yin, C., Gong, Y., & Pang, Y. (2003). Characterization of a chitin-binding protein GP37 of *Spodoptera litura* multicapsid nucleopolyhedrovirus. *Virus Research, 96,* 113-122.

McGuire, M. R., Tamez-Guerra, P., Behle, R. W., and Streett, D. A. (2001). Comparative field stability of selected entomopathogenic virus formulations. *Journal of Economic Entomology, 94,* 1037-1044.

Milks, M. L., Washburn, J. O., Willis, L. G., Volkman, L. E., & Theilmann, D. A. (2003). Deletion of *pe38* attenuates AcMNPV genome replication, budded virus production, and virulence in *Heliothis virescens*. *Virology, 310,* 224-234.

Mitsuhashi, W. (2002). Further evidence that spindles of an entomopoxvirus enhance its infectivity in a host insect. *Journal of Invertebrate Pathology, 79,* 59-61.

Mitsuhashi, W. (2006). Enhancement of the infectivity of insect viruses: a protein produced by entomopoxviruses is a potential synergist of viral pesticides. In S. G. Pandalai (Ed.), *Recent Research Developments in Entomology 5* (pp. 147-159), Kerala: Research Signpost.

Mitsuhashi, W., Furuta, Y., & Sato, M. (1998). The spindles of an entomopoxvirus of *Coleoptera* (*Anomala cuprea*) strongly enhance the infectivity of a nucleopolyhedrovirus in *Lepidoptera* (*Bombyx mori*). *Journal of Invertebrate Pathology, 71,* 186-188.

Mitsuhashi, W., Kawakita, H., Murakami, R., Takemoto, Y., Saiki, T., Miyamoto, K., & Wada, S. (2007). Spindles of an entomopoxvirus facilitate its infection of the host insect by disrupting the peritrophic membrane. *Journal of Virology, 81,* 4235-4243.

Mitsuhashi, W., & Miyamoto, K. (2003). Disintegration of the peritrophic membrane of silkworm larvae due to spindles of an entomopoxvirus. *Journal of Invertebrate Pathology, 82,* 34-40.

Mitsuhashi, W., & Murakami, R. (2007). Scanning electron microscopic observations of the peritrophic membrane in silkworm (*Bombyx mori*) larvae. *Transactions of the Lepidopterological Society of Japan, 58,* 172-176.

Mitsuhashi, W., Murakami, R., Takemoto, Y., Miyamoto, K., & Wada, S. (2008). Stability of the viral-enhancing ability of entomopoxvirus spindles exposed to various abiotic factors. *Applied Entomology and Zoology, 43,* 483-489.

Mitsuhashi, W., Saito, H., & Sato, M. (1997). Complete nucleotide sequence of the fusolin gene of an entomopoxvirus in the cupreous chafer, *Anomala cuprea* Hope (Coleoptera: Scarabaeidae). *Insect Biochemistry and Molecular Biology, 27,* 869-876.

Mitsuhashi, W., & Sato, M. (2000). Enhanced infection of a nucleopolyhedrovirus in a lepidopteran pest (*Spilosoma imparilis*) by spindles of a coleopteran entomopoxvirus (EPV) (*Anomala cuprea* EPV). *Journal of Forest Research, 5,* 285-287.

Mitsuhashi, W., Sato, M., & Hirai, Y. (2000). Involvement of spindles of an entomopoxvirus (EPV) in infectivity of the EPVs to their host insect. *Archives of Virology, 145,* 1465-1471.

Ohkawa, T., Washburn, J. O., Sitapara, R., Sid, E., & Volkman, L. E. (2005). Specific binding of *Autographa californica* M nucleopolyhedrovirus occlusion-derived virus to midgut cells of *Heliothis virescens* larvae is mediated by products of *pif* genes *Ac119* and *Ac022* but not by *Ac115*. *Journal of Virology, 79,* 15258-15264.

Okuno, S., Takatsuka, J., Nakai, M., Ototake, S., Masui, A., & Kunimi, Y. (2003). Viral-enhancing activity of various stilbene-derived brighteners for a *Spodoptera litura* (Lepidoptera: Noctuidae) nucleopolyhedrovirus. *Biological Control, 26,* 146-152.

Olszewski, J. A., & Dall, D. J. (2002). Assessment of foreign protein production by recombinant *Heliothis* (*Helicoverpa*) *armigera* entomopoxviruses in *Spodoptera frugiperda* cells. *Journal of General Virology, 83,* 451-461.

Peng, J., Zhong, J., & Granados, R. R. (1999). A baculovirus enhancin alters the permeability of a mucosal midgut peritrophic matrix from lepidopteran larvae. *Journal of Insect Physiology, 45,* 159-166.

Pijlman, G. P., Pruijssers, A. J. P., & Vlak, J. M. (2003). Identification of *pif-2*, a third conserved baculovirus gene required for *per os* infection of insects. *Journal of General Virology, 84,* 2041-2049.

Popham, H. J. R., Bischoff, D. S., & Slavicek, J. M. (2001). Both *Lymantria dispar* nucleopolyhedrovirus *enhancin* genes contribute to viral potency. *Journal of Virology, 75,* 8639-8648.

Rao, R., Fiandra, L., Giordana, B., Eguileor, M., Congiu, T., Burlini, N., Arciello, S., Corrado, G., & Pennacchio, F. (2004). AcMNPV ChiA protein disrupts the peritrophic membrane and alters midgut physiology of *Bombyx mori* larvae. *Insect Biochemistry and Molecular Biology, 34,* 1205-1213.

Roelvink, P. W., Corsaro, B. G., & Granados, R. R. (1995). Characterization of the *Helicoverpa armigera* and *Pseudaletia unipuncta* granulovirus enhancin genes. *Journal of General Virology, 76,* 2693-2705.

Shapiro, M., & Bell, R. A. (1982). Enhanced effectiveness of *Lymantria dispar* (Lepidoptera: Lymantriidae) nucleopolyhedrosis virus formulated with boric acid. *Annals of the Entomological Society of America, 75,* 346-349.

Shapiro, M., & Hamm, J. J. (1999). Enhancement in activity of homologous and heterologous baculoviruses infectious to fall armyworm (Lepidoptera: Noctuidae) by selected optical brighteners. *Journal of Entomological Science, 34,* 381-390.

Shapiro, M., Preisler, H. K., & Robertson, J. L. (1987). Enhancement of baculovirus activity on gypsy moth (Lepidoptera: Lymantriidae) by chitinase. *Journal of Economic Entomology, 80,* 1113-1116.

Shapiro, M., & Robertson, J. L. (1992). Enhancement of gypsy moth (Lepidoptera: Lymantriidae) baculovirus activity by optical brighteners. *Journal of Economic Entomology, 85,* 1120-1124.

Slack, J., & Arif, B. M. (2007). The baculoviruses occlusion-derived virus: virion structure and function. In K. Maramorosch & A. Shatkin. (Eds.), *Advances in Virus Research 69* (pp. 99-165), San Diego: Elsevier Inc.

Slack, J. M., Dougherty, E. M., & Lawrence, S. D. (2001). A study of the *Autographa californica* multiple nucleopolyhedrovirus ODV envelope protein p74 using a GFP tag. *Journal of General Virology, 82,* 2279-2287.

Slack, J. M., & Lawrence, S. D. (2005). Evidence for proteolytic cleavage of the baculovirus occlusion-derived virion envelope protein P74. *Journal of General Virology, 86,* 1637-1643.

Slavicek, J. M., & Popham, H. J. R. (2005). The *Lymantria dispar* nucleopolyhedrovirus enhancins are components of occlusion-derived virus. *Journal of Virology, 79,* 10578-10588.

Takemoto, Y., Mitsuhashi, W., Murakami, R., Konishi, H., & Miyamoto, K. (2008). The N-terminal region of an entomopoxvirus fusolin is essential for the enhancement of peroral infection, whereas the C-terminal region is eliminated in digestive juice. *Journal of Virology, 82,* 12406-12415.

Tanada, Y. (1959). Synergism between two viruses of the armyworm, *Pseudaletia unipuncta* (Haworth) (Lepidoptera, Noctuidae). *Journal of Insect Pathology, 1,* 215-231.

Tanada, Y. (1975). Enzyme synergistic for insect viruses. *Nature, 254,* 328-329.

Tanada, Y. (1985). A synopsis of studies on the synergistic property of an insect baculovirus: a tribute to Edward A. Steinhaus. *Journal of Invertebrate Pathology, 45,* 125-138.

Tanada, Y., Hess, R. T., Omi, E. M., & Yamamoto, T. (1983). Localization of a synergistic factor of a granulosis virus by its esterase activity in the larval midgut of the armyworm, *Pseudaletia unipuncta. Microbios, 37,* 87-93.

Tanada, Y., & Hukuhara, T. (1971). Enhanced infection of a nuclear-polyhedrosis virus in larvae of the armyworm, *Pseudaletia unipuncta*, by a factor in the capsule of a granulosis virus. *Journal of Invertebrate Pathology, 17,* 116-126.

Tanada, Y., & Kaya, H. K. (1993). *Insect Pathology*, San Diego: Academic Press Inc.

Tellam, R. L., Wijiffels, G., & Willadsen, P. (1999). Peritrophic matrix proteins. *Insect Biochemistry and Molecular Biology, 29,* 87-101.

Tsujibo, H., Orikoshi, H., Baba, N., Miyahara, M., Miyamoto, K., Yasuda, M., & Inamori, Y. (2002). Identification and characterization of the gene cluster involved in chitin degradation in a marine bacterium, *Alteromonas* sp. strain O-7. *Applied and Environmental Microbiology, 68,* 263-270.

Tuan, S., & Hou, R. F. (1988). Enhancement of nuclear polyhedrosis virus infection by lecithin in the corn earworm, *Heliothis armigera. Journal of Invertebrate Pathology, 52,* 180-182.

Vaaje-Kolstad, G., Horn, S. J., van Aalten, D. M. F., Synstad, B., & Eijsink, V. G. H. (2005). The non-catalytic chitin-binding protein CBP21 from *Serratia marcescens* is essential for chitin degradation. *The Journal of Biological Chemistry, 280,* 28492-28497.

Wang, P., & Granados, R. R. (1997). An intestinal mucin is the target substrate for a baculovirus enhancin. *Proceedings of the National Academy of Sciences, USA, 94,* 6977-6982.

Wang, P., & Granados, R. R. (1998). Observations on the presence of the peritrophic membrane in larval *Trichoplusia ni* and its role in limiting baculovirus infection. *Journal of Invertebrate Pathology, 72,* 57-62.

Wang, P., & Granados, R. R. (2000). Calcofluor disrupts the midgut defense system in insects. *Insect Biochemistry and Molecular Biology, 30,* 135-143.

Wang, P., & Granados, R. R. (2001). Molecular structure of the peritrophic membrane (PM): Identification of potential PM target sites for insect control. *Archives of Insect Biochemistry and Physiology, 47,* 110-118.

Wang, P., Hammer, D. A., & Granados., R. R. (1994). Interaction of *Trichoplusia ni* granulosis virus-encoded enhancin with the midgut epithelium and peritrophic membrane of four lepidopteran insects. *Journal of General Virology, 75,* 1961-1967.

Washburn, J. O., Chan, E. Y., Volkman, L. E., Aumiller, J. J., & Jarvis, D. L. (2003). Early synthesis of budded virus envelope fusion protein GP64 enhances *Autographa californica* multicapsid nucleopolyhedrovirus virulence in orally infected *Heliothis virescens*. *Journal of Virology, 77,* 280-290.

Webb, R. E., Dill, N. H., McLaughlin, J. M., Kershaw, L. S., Podgwaite, J. D., Cook, S. P., Thorpe, K. W., Farrar, R. R., Ridgway, R. L., Fuester, R. W., Shapiro, M., Argauer, R. J., Venables, L., & White, G. B. (1996). Blankophor BBH as an enhancer of nuclear polyhedrosis virus in arborist treatments against the gypsy moth (Lepidoptera: Lymantriidae). *Journal of Economic Entomology, 89,* 957-962.

Wijonarko, A., & Hukuhara, T. (1998). Detection of a virus enhancing factor in the spheroid, spindle, and virion of an entomopoxvirus. *Journal of Invertebrate Pathology, 72,* 82-86.

Xu, J., & Hukuhara, T. (1992). Enhanced infection of a nuclear polyhedrosis virus in larvae of the armyworm, *Pseudaletia separata*, by a factor in the spheroids of an entomopoxvirus. *Journal of Invertebrate Pathology, 60,* 259-264.

Xu, J., & Hukuhara, T. (1994). Biochemical properties of an enhancing factor of an entomopoxvirus. *Journal of Invertebrate Pathology, 63,* 14-18.

Yamamoto, T., & Tanada, Y. (1978a). Phospholipid, an enhancing component in the synergistic factor of a granulosis virus of the armyworm, *Pseudaletia unipuncta*. *Journal of Invertebrate Pathology, 31,* 48-56.

Yamamoto, T., & Tanada, Y. (1978b). Biochemical properties of viral envelopes of insect baculoviruses and their role in infectivity. *Journal of Invertebrate Pathology, 32,* 202-211.

Yao, L., Zhou, W., Xu, H., Zheng, Y., & Qi, Y. (2004). The *Heliothis armigera* single nucleocapsid nucleopolyhedrovirus envelope protein P74 is required for infection of the host midgut. *Virus Research, 104,* 111-121.

Zhang, J.-H., Ohkawa, T., Washburn, J. O., & Volkman, L. E. (2005). Effects of Ac150 on virulence and pathogenesis of *Autographa californica multiple nucleopolyhedrovirus* in noctuid hosts. *Journal of General Virology, 86,* 1619-1627.

Zhang, J.-H., Washburn, J. O., Jarvis, D. L., & Volkman, L. E. (2004). *Autographa californica* M nucleopolyhedrovirus early GP64 synthesis mitigates developmental resistance in orally infected noctuid hosts. *Journal of General Virology, 85,* 833-842.

Chapter 2

RNAi and the Study of Insect Immunity

E. P. Silva[1] and N. A. Ratcliffe[2]
[1]Instituto de Biologia, Universidade Federal Fluminense, Brazil
[2]Department of Biological Sciences, University of Swansea, UK

Abstract

RNA interference is a post-transcriptional mode of gene silencing that works as a cellular response to gene invasion resulting from transposons or viruses. Silencing occurs by a double-stranded RNA (homologous to the silenced gene) targeting and degrading the mRNA. To date, RNAi has been applied to a number of studies involved in insect immunity, such as the functional characterization of antimicrobial peptides and pattern recognition proteins, the encapsulation and phagocytosis processes, and the effects of viral infections and the modulation of the host immune response by insect parasitoids. These studies are reviewed in the present chapter and the importance of the application, as well as the usefulness of RNAi as a tool, are discussed in the light of the results obtained so far from studies on insect immunity.

Introduction

RNA interference (RNAi) is a post-transcriptional mode of gene silencing (Carthew, 2001) conserved in both plants and animals (Hannon, 2002). It is also one of the most important advances in biology in the last decades and has been developing rapidly over the past few years (Novina & Sharp, 2004). Its importance resides in the fact that it is a powerful tool for investigating gene function, especially with model organisms for which there is much information about the genome sequence. For other organisms, the application of RNAi can benefit from expressed sequence tag (EST) analysis or depends on the availability of cDNA sequences in the databases. It is particularly important in subjects such as genetics, molecular biology and physiology, and is also a potentially important tool in medical therapeutics (Mao et al., 2007).

Table 1. Different insect species to which RNAi technique has been applied

Species	Reference
Aedes aegypti	Bian et al., 2005; Cheon et al., 2006
Amblyomma americanum	Aljamali et al., 2003
Anopheles gambiae	Shiao et al., 2006
Antheraea pernyi	Hirai et al., 2004
Apis mellifera	Amdam et al., 2003; Weinstock et al., 2006
Apriona germari	Lee et al., 2006
Armigeres subalbatus	Wang et al., 2005
Blattella germânica	Martin et al., 2006; Cruz et al., 2007
Bombyx mori	Quan et al., 2002; Huang et al., 2007
Ceratitis capitata	Lamprou et al., 2007
Drosophila melanogaster	Korayem et al., 2004; Stroschein-Stevenson et al., 2006
Epiphyas postvittana	Turner et al., 2006
Gastrophysa atrocyanea	Tanaka & Suzuki, 2005
Glossina morsitans	Hu & Aksoy, 2006
Gryllus bimaculatus	Meyering-Vos et al., 2006
Harmonia axyridis	Kuwayama et al., 2006
Hyalophora cecropia	Bettencourt et al., 2002; Terenius et al., 2007
Mamestra brassicae	Tsuzuki et al., 2005
Manduca sexta	Vermehren et al., 2001; Eleftherianos et al., 2006[a]
Nasonia vitripennis	Lynch & Desplan, 2006
Plodia interpunctella	Fabrick et al., 2004
Pseudoplusia includens	Strand et al., 2006
Rhodnius prolixus	Araújo et al., 2006
Sarcophoga peregrina	Nishikawa & Natori, 2001
Spodoptera frugiperda	Meyering-Vos et al., 2006
Spodoptera litura	Rajagopal et al., 2002
Tenebrio molitor	Zhao et al., 2005
Tribolium castaneum	Bucher et al., 2002; Tomoyasu et al., 2008
Trichoplusia ni	Beck & Strand, 2003

As a research tool, RNAi has been applied widely, suggesting that it is a universal phenomenon in Eukaryotic organisms (Fire et al., 1998). To date, RNAi has been applied to a number of insects, including *Anopheles gambiae* (Shiao et al., 2006), *Antheraea pernyi* (Hirai et al., 2004), *Apis mellifera* (Amdam et al., 2003), *Blattella germanica* (Martin et al., 2006), *Bombyx mori* (Quan et al., 2002), *Gryllus bimaculatus* (Meyering-Vos et al., 2006), *Hyalophora cecropia* (Bettencourt et al., 2002), *Mamestra brassicae* (Tsuzuki et al., 2005), *Manduca sexta* (Levin et al., 2005; Vermehren et al., 2001, 2005), *Plodia interpunctella*

(Fabrick et al., 2004), *Rhodnius prolixus* (Araujo et al., 2006), *Sarcophoga peregrine* (Nishikawa & Natori, 2001) and *Spodoptera litura* (Rajagopal et al., 2002).

Despite the fact that RNAi has been applied to numerous different organisms, including several insect species, this technique has had limited success in some cases (Viney & Thompson, 2008). Some parasitic nematodes (Knox et al., 2007) have proven to be very difficult to knock-down, and many lepidopterans lack the ability to respond to dsRNA systemically (Marcus, 2005). Even *Drosophila*, the leading insect model organism, does not show a robust systemic RNAi response (Tomoyasu et al., 2008). The precise mechanisms involved in systemic RNAi in insects are still to be defined.

The innate immune system is a host defence mechanism that is evolutionarily conserved from plants to humans and is mainly involved in the recognition and control of early stage infection in animals (Hoffmann et al., 1999). Insect innate immunity shows a number of structural and functional similarities to the innate immune system of mammals (Vilmos & Kurucz, 1998) and it has been regarded in recent years as a model providing valuable insights into the basic mechanisms of the innate immune response in general (Lemaitre & Hoffmann, 2007). Also, insects are being developed as alternatives to vertebrates for assessing microbial virulence and determining the antimicrobial activity of drugs (Kavanagh & Reeves, 2004). RNAi is a particularly effective approach for researching the innate immune system (Mao et al., 2007) and has been applied to a number of studies on insect immunity (see Table 1). This review aims to examine the results of the latter work in order to assess the current progress in our understanding of insect immunity from the employment of RNAi. Initially, to assist the reader, a short overview is presented both of the RNAi process and of the general components of insect immunity.

RNA Interference

Plant biologists working with petunias were the first to observe the RNAi phenomenon. The goal of their work was to alter pigmentation by introducing exogenous transgenes into the plants. Surprisingly, the introduction of several copies of a gene that coded for a deep purple flower led to plants with white or patchy-coloured flowers. The introduced "transgenes" were causing silencing of both themselves and the plants' own colour genes (Napoli et al., 1990). The extent and importance of the phenomenon were confirmed by the studies with the worm *Caenorhabditis elegans*: the experimental introduction of double-stranded RNA (dsRNA) into cells was used to interfere with the function of an endogenous gene (Fire et al., 1998). Similar gene-silencing phenomena have, subsequently, been described for fungi, insects, mammals, crustaceans, etc. The generic term, RNA silencing, was coined to define these related RNA-guided gene regulatory mechanisms termed "post-transcriptional gene silencing" (PTGS) in plants, "quelling" in fungi and "RNAi" in animals (Ding, 2000).

RNAi-related pathways are an evolutionarily-conserved cellular response to gene invasion resulting from transposons or viruses (Lecellier & Voinnet, 2004). Since viruses can protect themselves by suppressing RNAi (Ding et al., 2004), interaction between RNA viruses and host RNAi represents a coevolutionary "arms race". Studies with *Drosophila* have shown that natural selection drives rapid evolution in genes related to RNAi which implies that coevolution between RNA viruses and host antiviral RNAi genes is active and significant in shaping RNAi function (Obbard et al., 2006).

The mechanism of RNAi involves the cleavage of cytoplasmic dsRNA, complementary to a portion of a gene, into 21–28 nucleotide duplexes called small interfering RNA (siRNA) by the RNase-III enzyme, Dicer. A single-strand of the siRNA duplex is then bound into the multiprotein RNA-inducing silencing complex (RISC) and used as a template to recognize and degrade the complementary endogenous messenger RNA (mRNA) (Fire, 2007). This process usually results in the inactivation of the target gene with the consequent phenotypic effect. One interesting feature of this process is the systemic effect of RNAi which starts with the injection of dsRNA into the animal (Hannon, 2002). Genes required for uptake and systemic spread of dsRNA have been identified. In *C. elegans*, *Sid-1* mutants cannot take-up or systemically spread externally supplied dsRNA. The *rsd-4* mutants have a similar phenotype and *Sid-2* mutants are also defective in the uptake but are competent to systemically spread RNAi introduced internally (Tijsterman et al., 2004; Winston et al., 2007). For insects, a *Sid-I* homologue (*AmSid-I*, Aronstein et al., 2006) was identified in the honey bee, *Apis melifera*, while work with *Drosophila* S2 cells indicates that dsRNA is taken up by an active process involving receptor-mediated endocytosis (Saleh et al., 2006). Although many metazoan cells can take up exogenous dsRNA and use it to initiate an RNA silencing response, the mechanism for this uptake is still ill-defined.

Insect Immunity

Insects are an example of evolutionary success being the single most abundant phylum on Earth. During evolution, insects developed an effective innate immune system which copes with pathogens such as viruses, bacteria, protozoans, fungi, nematodes and multicellular parasites and parasitoids (Hoffmann et al., 1999). Insect innate immunity results from three types of interacting mechanisms, namely, (i) mechanical barriers including the insect cuticle, (ii) humoral responses involving proteolytic cascades responsible for blood clotting, melanin formation, and opsonisation, as well as the synthesis of antimicrobial peptides (AMPs), and (iii) cellular reactions such as phagocytosis, nodule formation and encapsulation of invading organisms (Ratcliffe & Whitten, 2004).

The front line defence of insects is the resistant cuticle that rebuffs pathogens by forming a potent mechanical barrier. Moreover, there is evidence that the cuticle has additional pathogen resistance parameters. For example, genetics studies on the cuticle colouration of the mealworm, *Tenebrio molitor*, have demonstrated that haemocyte density and pre-immune challenge activity of phenoloxidase (PO) are significantly higher in black strains of beetles compared with tan strains (Armitage & Siva-Jothi, 2005).

Following an injury, a second line of defence is represented by the humoral reactions. The formation of the black pigment, melanin, is a rapid first response. Prophenoloxidase (PPO) is present in plasma and is activated by a serine protease cascade (Cerenius & Soderhall, 2004). After release by cell rupture, PPO is either actively transported into the cuticle or deposited around wounds (Ashida & Brey, 1995) and encapsulated parasites

(Bidochka et al., 1997). The activation of PPO can lead to the production of reactive oxygen species capable of killing sequestered invaders (Ratcliffe & Whitten, 2004). Haemolymph clotting is another mechanism activated by a serine proteolytic cascade. Serine protease cascades of insects have a dual role and are not only involved in hemolymph clotting and melanin formation, but are also toxic to microorganisms (Vilmos & Kurucz, 1998). A rapid and transient synthesis of a battery of AMPs is also an important part of the humoral insect defences. AMPs are produced mainly in the fat body, but the haemocytes, cuticular epithelial cells, the gut and salivary glands can produce AMPs (Ratcliffe & Whitten, 2004). These molecules are mostly of low molecular masses and cationic, exhibiting a broad spectrum of activity against bacteria and/or fungi (Lemaitre & Hoffmann, 2007; Hoffmann et al., 1996).

Regarding cellular reactions, only recently has anything been learnt about the molecular aspects of phagocytosis in insects. Granular cells and plasmatocytes are primarily responsible for phagocytosis (Ratcliffe & Whitten, 2004). Pathogens or parasites too numerous or too large to be dealt with by phagocytosis alone become enclosed in nodules and capsules. Nodules form by insect haemocytes aggregating to entrap large numbers of bacteria or fungi and then may be surrounded by layers of blood cells with the mature nodule eventually attaching to tissues. Encapsulation is usually a multicellular defence mechanism in which a capsule of overlapping layers of haemocytes is formed around an invading parasite (Ratcliffe, 1993; Vilmos & Kurucz, 1998). Pattern recognition proteins (PRP) that bind conserved pathogen-associated molecular pattern (PAMP) molecules produced by fungi and/or bacteria are important elements in the mediation of cellular responses (Fearon, 1997).

RNAi and Insect Immunity

Antimicrobial Peptides (AMPs)

The characterization of novel immune proteins is one field to which RNAi has been successfully applied. For example, study on AMPs in domesticated and wild silkmoths, *Bombyx mori* and *Antheraea mylitta*, led to identification of a new class of AMPs. These AMPs were designated as lysozyme-like proteins (LLPs) owing to their partial similarity with lysozymes. Further investigation of their antibacterial mechanisms has demonstrated that one of them (*Bombyx mori* LLP, BLLP1) is bacteriostatic rather than bactericidal against *Escherichia coli* and *Micrococcus luteus*. RNAi knock-down of BLLP1 resulted in a substantial bacterial load increase in the haemolymph (Gandhe et al., 2007). It was demonstrated that the antibacterial mechanism of this protein depends on peptidoglycan binding unlike peptidoglycan hydrolysis or membrane permeabilization as observed with lysozymes and most other antimicrobial peptides. This was one of the first reports on functional analysis of novel, noncatalytic lysozyme-like family of antibacterial proteins that are quite apart functionally from classical lysozymes.

One of the two major pathways regulating innate immunity in invertebrates is the immunodefiency (Imd) pathway (Hoffmann & Reichhart, 2002). The Imd signaling cascade is similar to the mammalian TNF-receptor pathway and has been shown, in *Drosophila* and other insects, to control AMP expression against Gram-negative bacteria (De Gregorio et al., 2002). Large-scale RNAi screening has been used to identify novel components of the Imd pathway in *Drosophila* S2 cells. In total, 6713 dsRNAs from an S2 cell-derived cDNA library were analyzed for their effect on the attacin promoter activity in response to *E. coli*. From this total, two novel Imd pathway components were identified, namely, an inhibitor of apoptosis 2 (Iap2) and a transforming growth factor-activated kinase 1 (TAK1)-binding protein (TAB) (Kleino et al., 2005). Further genome-wide kinetic analysis of the role of TAK1-TAB2 and Iap2 in the immune response of *Drosophila* S2 cells, using oligonucleotide microarrays, showed that TAB2 RNAi abolished the induction of all immune response genes in S2 cells, indicating its requirement for signaling via both the Imd (imuno deficiency) and the JNK (c-Jun N-terminal Kinase) pathways. The role of Iap2 was more specific since kinetic analysis indicated that Iap2 is required to sustain antimicrobial peptide gene expression in S2 cells. Furthermore, inactivation of Iap2 by RNAi resulted in impaired microbial resistance in *Drosophila in vivo* (Valanne et al., 2007).

The Imd pathway was also investigated in tsetse flies, *Glossina morsitans morsitans*, by RNAi. For this species, the molecular structure of the Imd pathway transcriptional activator Relish (GmmRel) shows high amino acid identity and structural similarity to it's *Drosophila* homologue (Hu & Aksoy, 2006). Tsetse flies transmit African trypanosomes, which are the protozoan parasites and agents of human sleeping sickness in sub-Saharan Africa. In a study by Hu & Aksoy (2006), the focus was on the tsetse fly's natural defence against *Trypanosoma brucei* infections. Through an RNAi approach it was shown that the pathogen-induced expression profile of the AMPs, *attacin* and *cecropin,* is under the regulation of GmmRel. Knockdown of *GmmRel* successfully blocked the induction of *attacin* and *cecropin* expression in the immune responsive tissues, the fat body and proventriculus (cardia), following microbial challenge. The trypanosome infection rates in the midgut and salivary glands, as well as the density of midgut parasite infections were found to be significantly higher in flies when *attacin* and *relish* expression were knocked down (Hu & Aksoy, 2006).

Toll receptors, the second major pathway involved in intracellular signal transduction and initiation of insect antimicrobial immune responses, have also been studied with the help of RNAi. For example, Kambris et al. (2006) have used an *in vivo* RNAi strategy to inactivate *Drosophila* serine protease (SR) genes. Toll receptor in *Drosophila* does not interact directly with microbial determinants but is rather activated upon binding a cleaved form of the cytokine-like molecule Spatzle (Spz). During the immune response, Spz is thought to be processed by secreted SPs present in the hemolymph that are activated by the recognition of gram-positive bacteria or fungi. Their screen of a collection of 75 distinct SP genes regulating the activation of the Toll pathway by gram-positive bacteria identified five novel SPs that function in an extracellular pathway linking the recognition proteins GNBP1 (gram-negative bacteria binding protein 1) and PGRP-AS (peptidoglycan recognition protein SA) to Spz. Four of these genes were also required for Toll activation by fungi, while one was specifically associated with signaling in response to gram-positive bacterial infections. These results demonstrated the existence of a common cascade of SPs upstream of Spz, integrating signals sent by various secreted recognition molecules via more specialized SPs. The function of a novel gene (Am18w) from the honey bee, *Apis mellifera*, which encodes for a Toll-like receptor and shares similarity with *Bombyx mori* 18-wheeler (51.4%),

Drosophila Toll-7 receptor (46.6%) and *Drosophila* 18-wheeler (42.5%), was investigated using injections of homologous dsRNA which successfully disrupted the endogenous mRNA of the target gene (Aronstein & Saldivar, 2005). It was demonstrated that expression levels of AMPs were reduced with the knock-down of this Am18w gene. Similarly, to elucidate the role of mosquito REL1 (AaREL1), a homolog of *Drosophila* Dorsal, in regulation of the Toll immune pathway in the mosquito, *Aedes aegypti*, Bian et al. (2005) took the transgenic approach, in combination with RNAi. The effect of the transgenic RNAi knockdown of AaREL1 on mosquito innate immunity was revealed by increased susceptibility to the entomopathogenic fungus, *Beauveria bassiana* and the reduced induction of Spz1A and Serpin-27A gene expressions after fungal challenge. These results have proven that AaREL1 is a key downstream regulator of the Toll immune pathway in the mosquito, *A. aegypti*. Taken together, the results from *Drosophila*, *Glossina*, *Apis* and *Aedes* show that the RNAi-based approach is suitable for identifying and characterizing genes in conserved signaling cascades.

Melanization and Clotting

Melanization and clotting are critical mechanisms in initiating wound healing and limiting haemolymph loss in insects. They are also important immune defences, quickly forming a secondary barrier to infection, immobilizing bacteria and promoting their killing (Theopold et al., 2004; Strand, 2008). In *Drosophila,* many genes involved with melanization and clotting has been characterized with the help of RNAi. For example, Fondue (fon; CG15825), an immune-responsive gene of *Drosophila melanogaster* that encodes an abundant haemolymph protein containing multiple repeat blocks was shown to be required for efficient clotting of the blood (Scherfer et al., 2006). In addition, RNAi showed that gp150, an ecdysone-regulated mucin that is found in haemocytes, the gut (peritrophic membrane) and in the salivary glands, was involved as part of the clot and participated in the entrapment of bacteria (Korayem et al., 2004). Furthermore, silencing of the gene Sp7 (a serine protease) by RNAi impaired the melanization reaction upon injury, thus demonstrating that Sp7 is required for PO activation (Castillejo-López & Häcker, 2005). Also, hemolectin (Hml) RNAi-treated *Drosophila* larvae show a bleeding defect upon injury with failure to seal a wound completely, strongly suggesting that Hml is involved in haemostasis and/or coagulation (Goto et al., 2003).

The melanization reaction induced by activated PO was studied in Lepidopteran and beetles. Terenius et al. (2007) injected dsRNA of the lepidopteran immune protein hemolin in pupae of *Hyalophora cecropia*. They observed a significant reduction in PO activity after 24 h, but not after 72 h. The link between hemolin and the PPO system suggests that hemolin can be working as pattern recognition protein important for the triggering of the PPO cascade in the defense against bacterial infections. Because excessive formation of quinones and systemic hypermelanization are deleterious to the hosts, Zhao et al. (2005) studied the melanization process in the beetle *Tenebrio molitor* especially interested in the control mechanisms involved in PPO reaction. They reported the purification and cloning of a cDNA of a novel 43-kDa protein, which functions as a melanization-inhibiting protein (MIP). RNAi using a synthetic 445-mer double-stranded RNA of MIP injected into *Tenebrio* larvae showed that melanin synthesis was markedly induced. Their results strongly suggest that this 43-kDa MIP inhibits the formation of melanin and thus is a modulator of the melanization reaction to prevent the insect from excessive melanin synthesis in places where it should be inappropriate.

Pattern Recognition Proteins

Pattern recognition proteins (PRP) in insects bind to molecular patterns present on the surface of pathogens, such as bacteria and fungi, and trigger a protective response involving humoral and cellular reactions. Cellular mechanisms mediated by haemocytes include phagocytosis, encapsulation and the formation of melanotic nodules. RNAi has been used as a tool to better understand these processes.

Eleftherianos et al. (2006a) showed that two different species of insect pathogenic bacteria, *Photorhabdus luminescens* TT01 and *Photorhabdus asymbiotica* ATCC43949, are both recognized by the immune system of their host insect, the tobacco hornworm, *Manduca sexta*, as indicated by a rapid increase in the levels of mRNAs encoding three different inducible microbial recognition proteins, namely, hemolin, immulectin-2 and peptidoglycan recognition protein (PGRP). Knock-down of any one of these genes markedly decreased the ability of the insects to withstand infection when exposed to either species of *Photorhabdus*, as measured by the rate at which infected insects died. RNAi against immulectin-2 caused the greatest reduction in host resistance to infection. The decreased resistance to infection was associated with reduced haemolymph PO activity. Eleftherianos et al. (2007) have also shown with *Manduca* that a non-pathogenic strain of *E. coli* induces mRNA transcription and protein expression of hemolin and PGRP but not immulectin-2 in the haemocytes. However, it is the knock-down of hemolin that significantly decreased the ability of insects to clear *E. coli* from the haemolymph and caused a reduction in the number of free haemocytes. Moreover, RNAi of hemolin reduced the ability of haemocytes to engulf bacteria through phagocytosis and to form melanotic nodules *in vivo*. They also demonstrated that washed haemocytes taken from RNAi-treated insects showed reduced ability to form microaggregates around bacteria *in vitro* showing that the immune function affected by RNAi knockdown of hemolin is intrinsic to the haemocytes. Taken together, these results demonstrate that *Manduca sexta* immune system is able to recognize infection promoted by *Photorhabdus* and *E. coli* and that this recognition is important in triggering the cellular immune responses. Very interesting also is the immune "memory" presented by *M. sexta* caterpillars which elicits effective immunity against infection by *Photorhabdus luminescens* TT01 after having prior infection with the non-pathogenic bacterium *E. coli* (Eleftherianos et al., 2006b). Induction of this protective effect is associated with up-regulation of both the microbial pattern recognition genes (hemolin, immulectin-2 and PGRP) as well as the anti-bacterial effector genes (AMPs, attacin, cecropin, lebocin, lysozyme and moricin). RNAi knockdown of expression of individual recognition proteins had a more drastic adverse effect on the *E. coli*-elicited immunity than interfering with expression of individual AMPs. RNAi knock-down of immulectin-2 caused the greatest reduction in immunity, followed by hemolin and

PGRP. These results demonstrate that the insect immune system can be effectively primed by prior infection with nonpathogenic bacteria against subsequent infection by a highly virulent pathogen.

As shown by the work of Eleftherianos and collaborators (2006a, 2006b, 2007) with *M. sexta*, hemolin is a protein strongly induced upon bacterial infection and using RNAi also unequivocally confirmed previous results that it plays an important role as a PRP. One previous study of hemolin was that of Hirai et al. (2004) which showed that when they applied dsRNA to suppress hemolin expression in the Chinese oak silk moth, *Antheraea pernyi*, hemolin was induced rather than knocked down by the RNAi treatment. They explained their result as if dsRNA was recognized as a virus pattern molecule. Further investigating the problem, they used the baculovirus (*Ap* NPV) to infect the insects and found that hemolin was induced and expressed with similar kinetics to that following dsRNA injection. No subsequent report appeared of such an effect of dsRNA on hemolin which makes the Hirai et al. (2004) result very intriguing.

RNAi is also an important tool for the characterization of novel PRPs and for discriminating the functions of different PRP isoforms. For example, characterization of a novel haemocyte-specific β,1,3-glucan recognition protein (GRP) from the mosquito, *Armigeres subalbatus* (AsGRP), was undertaken with the help of the RNAi technique (Wang et al., 2005). GRPs have specific affinity for β, 1,3-glucans, a component on the surface of fungi and bacteria. By interacting with β, 1,3-glucan, GRP initiates activation of PPO. As GRP is constitutively expressed in the haemolymph of adult female *A. subalbatus*, and is upregulated following challenge with *E. coli*, *Micrococcus luteus*, and the filarial worm *Dirofilaria immitis*. RNAi against AsGRP strongly inhibits melanotic encapsulation of *D. immitis*. Besides the demonstration that AsGRP is involved in the encapsulation of *D. immitis* in *A. subalbatus*, this result also suggests that AsGRP possibly has the capacity to bind to a variety of pathogens, functioning as a PRP. RNAi has also been used for discriminating the functions of peptidoglycan recognition protein (PGRP-LC) isoforms in *Drosophila* (Werner et al., 2003). The PGRP-LC is a major activator of the *imd/Relish* pathway in the *Drosophila* immune response. Three transcripts are generated by alternative splicing of the complex PGRP-LC gene. The encoded transmembrane proteins share an identical intracellular domain, but each has a separate extracellular PGRP-domain, namely, *x*, *y*, or *a*. Using RNA interference in *Drosophila* mbn-2 cells, Werner et al. (2003) showed that PGRP-LCx is the only isoform required to mediate signals from Gram-positive bacteria and purified bacterial peptidoglycan. In contrast, the recognition of Gram-negative bacteria and bacterial lipopolysaccharide requires both PGRP-LCa and LCx. The third isoform, LCy, was expressed at lower levels and possibly is partially redundant. These results from *A. subalbatus* and *Drosophila* are good examples of the power of RNAi as a tool to enlighten the sometimes grey zones of the pathways in the innate immune system of insects.

Levin et al. (2005) reported the identification of an integrin present exclusively on the surface of hemocytes in the tobacco hornworm, *Manduca sexta*. Monoclonal antibodies MS13 and MS34, which bind to plasmatocytes and block encapsulation, demonstrated in immunoaffinity chromatography experiments that this integrin was the same integrin b subunit. A cDNA was cloned and characterized. Injection of double stranded integrin-b1 RNA into larvae resulted in decreased integrin b1 expression in plasmatocytes and significantly suppressed encapsulation. These results indicated that activation of ligand-binding by the hemocyte-specific integrin plays a key role in stimulating plasmatocyte adhesion leading to encapsulation.

Phagocytosis

Phagocytosis is an important innate immune response against pathogens and parasites, and is a highly conserved aspect of innate immunity from insects to humans. Rämet et al. (2002) devised an RNAi-based screen in macrophage-like *Drosophila* S2 cells, and defined 34 gene products involved in phagocytosis. These include proteins that participate in haemocyte development, vesicle transport, actin cytoskeleton regulation and a cell surface receptor. This receptor, peptidoglycan recognition protein LC (PGRP-LC), was shown to be involved in recognition and signaling of Gram-negative bacteria. Similarly, Stroschein-Stevenson et al. (2006) used *Drosophila melanogaster* S2 cells as a model system to study the phagocytosis of *Candida albicans*, by screening an RNAi library representing 7,216 fly genes conserved among metazoans. After rescreening the initial genes identified and eliminating certain classes of housekeeping genes, they identified 184 genes required for efficient phagocytosis of *C. albicans* by the S2 cells. Diverse biological processes were represented, with actin cytoskeleton regulation, vesicle transport, signaling, and transcriptional regulation being prominent. Secondary screens, using *E. coli* and latex beads, revealed several genes specific for *C. albicans* phagocytosis. Characterization of one of these gene products, macroglobulin complement related (Mcr) protein, showed that it is secreted, binds specifically to the surface of *C. albicans*, and that it promotes subsequent phagocytosis.

Mcr is closely related to the four *Drosophila* thioester proteins (Teps). Stroschein-Stevenson et al. (2006) showed that TepII is required for efficient phagocytosis of *E. coli* (but not *C. albicans* or *Staphylococcus aureus*) and that TepIII is required for the efficient phagocytosis of *S. aureus* (but not *C. albicans* or *E. coli*). Thus, these families (Teps and Mcr) of fly proteins distinguish different pathogens for subsequent phagocytosis. Still in *Drosophila*, Philips et al. (2005) using a genome-wide RNAi screen in macrophage-like cells, using *Mycobacterium fortuitum* were able to identify factors required for general phagocytosis, as well as those needed specifically for mycobacterial infection. One of these factors was the so called Peste (Pes), which is a CD36 family member required for uptake of *Mycobacteria*, but not *Escherichia coli* or *Staphylococcus aureus*. These results show that involved in phagocytosis process are many genes and their products in a series of complex interactions. Phagocytosis is an important innate immune response against pathogens and parasites, and several signal transduction pathways regulate this process. Generally, RNAi approach applied to phagocytosis phenomena involves the screening of a large number of genes in model animals or their cells.

Alternatively to the large scale screen approach Kurucz *et al.* (2007) were able to identify and demonstrate the involvement of a protein, Nimrod C1 (NimC1), which is involved in the phagocytosis of bacteria. NimC1 is a 90–100 kDa single-pass transmembrane protein with ten characteristic EGF-like repeats (NIM repeats). It can be found on the surface of circulating and sessile plasmatocytes. The NimC1 gene is part of a cluster of ten related nimrod genes at 34E on chromosome 2, and similar clusters of nimrod-like genes are conserved in other insects such as *Anopheles* and *Apis*. RNAi Suppression of NimC1 expression in plasmatocytes inhibited the phagocytosis of *Staphylococcus aureus*. Conversely, overexpression of NimC1 in S2 cells stimulated the phagocytosis of both *S. aureus* and *E. coli*. Most interstingly, the Nimrod proteins are related to other putative phagocytosis receptors such as Eater and Draper from *D. melanogaster* and CED-1 from *C. elegans*. Together, they form a superfamily that also includes proteins that are encoded in the human genome. In the same way, in one of the first RNAi reports in insects, Nishikawa & Natori (2001) targeted disruption of a pupal hemocyte protein of *Sarcophaga* by RNA interference. This protein was previously purified and identified as a transmembrane protein with a molecular mass of 120 kDa (p120) that was exclusively expressed in pupal hemocytes. Double-stranded RNA injected into the larval body cavity effectively inhibited the expression of p120 in pupal hemocytes which were found to have lost the ability to take up acetylated low density lipoprotein, indicating that p120 is a scavenger receptor specifically expressed on the surface of pupal hemocytes.

Parasitoids and Viruses

Engineered viruses (modified O´nyong-nyong virus, ONNV, Keene et al., 2004) and cells (loss of function mutation, Galiana-Arnoux et al., 2006) associated with RNAi technique were used to study the importance of RNA interference for controlling viral replication *in vivo* and also to identify locus which are involved in host susceptibility to viral infection and spreading in model organisms such as *Drosophila* (Galiana-Arnoux et al., 2006) and mosquitoes (Keene et al., 2004). Complementarily RNAi has been effectively employed to confirm the functional roles of proteins involved in the insect immune response to viruses and the eggs and larvae of parasitic wasp. Some viruses are symbiotically associated with certain parasitoid wasps. This is the case with the braconid wasp, *Microplitis demolitor*, which carries the *M. demolitor* bracovirus (MdBV), a member of the Polydnaviridae family of dsDNA viruses, and parasitizes the larval stages of several noctuid moths. Beck & Strand (2003) used RNAi to specifically silence the glc1.8 and egf1.0 genes from MdBV in High Five cells derived from the lepidopteran, *Trichoplusia ni*. Dose-response studies indicated that MdBV infects High Five cells and blocks the ability of these cells to adhere to culture plates. When complementary dsRNA to glc1.8 was used, there was successful silencing of this gene and restoration of the adhesive phenotype of High Five cells. In contrast, when egf1.0 was successfully silenced there was no effect on adhesion. This result demonstrated that glc1.8 from MdBV has an important role in the disruption of adhesion in infected host cells. Similarly, RNAi was used by Strand et al. (2006) to determine which haemocyte types were phagocytic in the lepidopteran, *Pseudoplusia includens,* and to assess whether MdBV infection affects this defence response. They used the bacterium *E. coli* and inert polystyrene beads as targets and knocked-down the glc1.8 gene. Their results indicated that the professional phagocytes in *P. includens* were granulocytes and, once more, showed that glc1.8 was a key virulence determinant in disruption of both adhesion and phagocytosis by the insect immune cells.

Refractoriness

Despite the fact that RNAi has been sucesfully applied in several different fields of insect immunity investigation (as discussed above), there is increasing evidence that the success of the RNAi technique can be highly variable with some insects (Tomoyasu et al., 2008) and/or genes (the authors of this chapter have performed exhaustive studies on immune-related genes from the lepidopteran *Galleria mellonella* which have shown refractoriness to knockdown; paper in preparation). The literature on RNAi indicates that the efficiency of RNAi in the knockdown process can vary extensively (Amdam et al., 2003; Fabrick et al., 2004; Araujo et al., 2006). Tomoyasu et al. (2008) discuss the fact that some insects do not show a robust systemic RNAi response. These include the silkmoth, *Bombyx mori*, and *Drosophila*, in contrast to other insects such as *Tribolium castaneum* with a highly effective RNAi response. The possible reason for this variation may well be related to differences in the composition of component genes involved in the RNAi pathway, such as *sid-1*, *rsd-3*, and *Dcr-1* (Tomoyasu et al., 2008). A description of the cellular machinery of RNAi will be helpful in understanding the sucesses and failures of RNAi as applied to the study of insect immunity (Reynolds & Eleftherianos, 2008). An alternative possibility for this variation in RNAi efficiency could also be the acquisition of an FHV virus infection in the insect stocks. Such infections would also explain why even with the same insect species variations have been found in RNAi efficiency (e.g., *Bombyx* refered in Tomoyasu et al., 2008).

Conclusion

RNAi probably first evolved as an innate immune mechanism against viruses; however, it stands nowadays as the most powerful methodology to clarify many of the mechanisms involved in innate immunity. Among the insects, *Drosophila*, with its powerful genetics, has proven especially important for the study of humoral innate immunity with two signaling pathways—namely, Toll and Imd—involved. Also, *Drosophila* cells have been used for investigating the cellular immune responses, focusing on the mechanisms of phagocytosis, antiviral immunity and viral pathogenesis (e.g., Foley & O'Farrel, 2004; Cherry & Silverman, 2006; Wang & Ligoxygakis, 2006). These studies have taken advantage of genome-wide RNAi screening in *Drosophila* cells. Other insects that have been in the forefront of functional genomic approaches are the mosquitoes (Shin et al., 2003; Osta et al., 2004) for which research has been advanced by several breakthroughs, including the successful transformation of parasite vectors, DNA microarrays and, most importantly, genome sequencing, as in the case of *Anopheles gambiae*. In this review, the focus was mainly on non-*Drosophila* insect species less studied so far by the RNAi approach. The examples offered here make it clear that RNAi technology is a powerful tool applicable, with difficulty in some cases, to the field of insect immunity where it can help to elucidate numerous mechanisms in the innate immune response.

References

Aljamali, M.N.; Bior, A.D.; Sauer, J.R. & Essenberg, R.C. 2003. RNA interference in ticks: a study using histamine binding protein dsRNA in the female tick *Amblyomma americanum*. *Insect Molecular Biology* 12(3):299–305.

Amdam, G.V.; Simões, Z.L.P.; Guidugli, K.R.; Norberg, K. & Omholt, S.W. 2003. Disruption of vitellogenin gene function in adult honeybees by intra-abdominal injection of double-stranded RNA. *BMC Biotechnology* 3:8 p.

Araujo, R.N.; Santos, A.; Pinto, F.S.; Gontijo, N.S.; Lehane, M.J. & Pereira, M.H. 2006. RNA interference of the salivary gland nitrophorin 2 in the triatomine bug *Rhodnius prolixus* (Hemiptera: Reduviidae) by dsRNA ingestion or injection. *Insect Biochemistry and Molecular Biology* 36:683-693

Armitage, S.A.O. & Siva-Jothi, M.T. 2005. Immune function responds to selection for cuticular colour in *Tenebrio molitor*. *Heredity* 94(6):650-656.

Aronstein, K. & Saldivar, E. 2005. Characterization of a honey bee Toll related receptor gene Am18w and its potential involvement in antimicrobial immune defense. *Apidologie* 36(1):3-14.

Aronstein, K; Pankiw, T. & Saldivar, E. 2006. SID-I is implicated in systemic gene silencing in the honey bee. *Journal of Apicultural Research* 45(1):20-24.

Ashida, M. & Brey, P.T. 1995. Role of the integument in insect defence: Pro-phenol oxidase cascade in the cuticular matrix. *Proc. Natl. Acad. Sci. USA* 92(23):10698-10702.

Beck, M. & Strand, M.R. 2003. RNA interference silences *Microplitis demolitor* bracovirus genes and implicates glc1.8 in disruption of adhesion in infected host cells. *Virology* 314(2):521-535.

Bettencourt, R.; Terenius, O. & Faye, I. 2002. *Hemolin* gene silencing by ds-RNA injected into *Cecropia* pupae is lethal to next generation embryos. *Insect Molecular Biology* 11(3):267-271.

Bian, G; Shin, S.W.; Cheon, H.M.; Kokoza, V. & Raikhel, A.S. 2005. Transgenic alteration of Toll immune pathway in the female mosquito *Aedes aegypti*. *Proc. Natl. Acad. Sci. USA* 102(38):13568-13573.

Bidochka, M.J.; St. Leger, R.J. & Roberts, D.W. 1997. Induction of novel proteins in *Manduca sexta* and *Blaberus giganteus* as a response to fungal challenge. *Journal of Invertebrate Pathology* 70:184-189.

Bucher, G.; Scholten, J. & Klingler, M. 2002. Parental RNAi in *Tribolium* (Coleóptera). *Curr. Biol.* 12:R85-R86.

Carthew, R.W. 2001. Gene silencing by double-stranded RNA. *Current Opinion in Cell Biology* 13:244-248.

Castillejo-López, C. & Hacker, U. 2005. The serine protease Sp7 is expressed in blood cells and regulates the melanization reaction in *Drosophila*. *Biochemical and Biophysical Research Communications* 338(2):1075-1082.

Cerenius, L. & Soderhall, K. 2004. The prophenoloxidase-activating system in invertebrates. *Immunol. Rev.* 198: 116-126.

Cheon, H.-M.; Shin, S.W.; Bian, G.; Park, J.-H. & Raikhel, A.S. 2006. Regulation of lipid metabolism genes, lipid carrier protein lipophorin, and its receptor during immune challenge in the mosquito *Aedes aegypti*. *J. Biological Chemistry* 281(13):8426–8435.

Cherry, S. & Silverman, N. 2006. Host-pathogen interactions in drosophila: new tricks from an old friend. *Nature Immunology* 7(9):911-917.

Cruz, J.; Martın, D. & Belles, X. 2007. Redundant ecdysis regulatory functions of three nuclear receptor HR3 isoforms in the direct-developing insect *Blattella germânica*. *Mechanisms of Development* 124:180–189.

De Gregorio, E.; Spellman, P.T.; Tzou, P.; Rubin, G.M. & Lemaitre, B. 2002. The Toll and Imd pathways are the major regulators of the immune response in *Drosophila*. *EMBO Journal* 21(11):2568-2579.

Ding, S.W. 2000. RNA silencing. *Current Opinion in Biotechnology* 11:152-156.

Ding, S.W.; Li, H.W.; Lu, R.; Li, F. & Li, W.X. 2004. RNA silencing: a conserved antiviral immunity of plants and animals. *Virus Research* 102(1):109-115.

Ehlers, D.; Zosel, B.; Mohrig, W.; Kauschle, E. & Ehlers, D. 1992. Comparison of *in vivo* and *in vitro* phagocytosis in *Galleria mellonella* L. *Parasitol. Res.* 78: 354-359.

Eleftherianos, I.; Marokhazi, J.; Millichap, P.J.; Hodgkinson, A.J.; Sriboonlert, A.; Ffrench-Constant, R.H. & Reynolds, S.E. 2006a. Prior infection of *Manduca sexta* with non-pathogenic *Escherichia coli* elicits immunity to pathogenic *Photorhabdus luminescens*: Roles of immune-related proteins shown by RNA interference. *Insect Biochemistry and Molecular Biology* 36:517-525.

Eleftherianos, I.; Millichap, P.J.; Ffrench-Constant, R.H. & Reynolds, S.E. 2006b. RNAi suppression of recognition protein mediated immune responses in the tabacco hornworm *Manduca sexta* causes increased susceptibility to the insect pathogen *Photorhabdus*. *Developmental and Comparative Immunology* 30:1099-1107.

Eleftherianos, I.; Gokcen, F.; Felfoldi, G.; Millichap, P.J.; Trenczek, T.E.; Ffrench-Constant, R.H. & Reynolds, S.E. 2007. The immunoglobulin family protein Hemolin mediates cellular immune responses to bacteria in the insect *Manduca sexta*. *Cellular Microbiology* 9(5):1137-1147.

Fabrick, J.; Kanost, M.R. & Baker, J.E. 2004. RNAi-induced silencing of embryonic tryptophan oxygenase in the Pyralid moth *Plodia interpunctella*. *Journal of Insect Science* 4:1-9.

Fearon, D.T. 1997. Seeking wisdom in innate immunity. *Nature* 388:323-324.

Fire, A.Z. 2007. Gene Silencing by Double-Stranded RNA (Nobel Lecture). *Angew. Chem. Int. Ed.* 46:6966–6984.

Fire, A.; Xu, S.Q.; Montgomery, M.K.; Kostas, S.A.; Driver, S.E. & Mello, C.C. 1998. Potent and specific genetic interference by double-stranded RNA in *Caenorhabditis elegans*. *Nature* 391:806-811.

Foley, E. & O'Farrell, P.H. 2004. Functional dissection of an innate immune response by a genome-wide RNAi screen. *PLOS Biology* 2(8):1091-1106.

Galiana-Arnoux, D.; Dostert, C.; Schneemann, A.; Hoffmann, J.A. & Imler, J.L. 2006. Essential function *in vivo* for Dicer-2 in host defense against RNA viruses in drosophila. *Nature Immunology* 7(6):590-597.

Gandhe, A.S.; Janardhan, G. & Nagaraju, J. 2007. Immune upregulation of novel antibacterial proteins from silkmoths (Lepidoptera) that resemble lysozymes but lack muramidase activity. *Insect Biochemistry and Molecular Biology* 37(7):655-666.

Goto, A.; Kadowaki, T. & Kitagawa, Y. 2003. *Drosophila* hemolectin gene is expressed in embryonic and larval hemocytes and its knock down causes bleeding defects. *Developmental Biology* 264(2):582-591.

Hannon, G.J. 2002. RNA interference. *Nature* 418:244-251.

Hirai, M.; Terenius, O. & Faye, I. 2004. Baculovirus and dsRNA induce Hemolin, but no antibacterial activity, in *Antheraea pernyi*. *Insect Molecular Biology* 13(4):399-405.

Hoffmann, J.A. & Reichhart, J.-M. 2002. *Drosophila* innate immunity: an evolutionary perspective. *Nature Imunology* 3(2):121-126.

Hoffmann, J.A.; Reichhart, J.-M. & Hetru, C. 1996. Innate immunity in higher insects. *Current Opinion in Immunology* 8:8-13.

Hoffmann J.A.; Kafatos, F.C.; Janeway, C.A. & Ezekowitz, R.A.B. 1999. Phylogenetic perspectives in innate immunity. *Science* 284:1313-1318.

Hu, C.Y. & Aksoy, S. 2006. Innate immune responses regulate trypanosome parasite infection of the tsetse fly *Glossina morsitans morsitans*. *Molecular Microbiology* 60(5):1194-1204.

Huang, J.; Zhang, Y.; Li, M.; Wang, S.; Liu, W.; Couble, P.; Zhan, G. & Huang, Y. 2007. RNA interference-mediated silence of the bursicon gene induces defects in wing expansion in silkworm. *FEBS Letters* 581:697-701.

Kambris, Z.; Brun, S.; Jang, I.-H.; Nam, H.-J.; Takahashi, K.; Lee, W.-J.; Ueda, R. & Lemaitre, B. 2006. *Drosophila* immunity: A large scale *in vivo* RNAi screen identifies five serine proteases required for Toll activation. *Curr. Biol.* 16: 808-813.

Kavanagh, K. & Reeves, E.P. 2004. Exploiting the potential of insects for in vivo pathogenicity testing of microbial pathogens. *FEMS Microbiology Reviews* 28:101-112.

Keene, K.M.; Foy, B.D.; Sanchez-Vargas, I.; Beaty, B.J.; Blair, C.D. & Olson, K.E. 2004. RNA interference acts as a natural antiviral response to O'nyong-nyong virus (Alphavirus; Togaviridae) infection of *Anopheles gambiae*. *Proc. Natl. Acad. Sci. USA* 101(49):17240-17245.

Kleino, A.; Valanne, S.; Ulvila, J.; Kallio, J.; Myllymaki, H.; Enwald, H.; Stoven, S.; Poidevin, M.; Ueda, R.; Hultmark, D.; Lemaitre, B. & Ramet, M. 2005. Inhibitor of apoptosis 2 and TAK1-binding protein are components of the *Drosophila* Imd pathway. *EMBO Journal* 24(19):3423-3434.

Knox, D.P.; Geldhof, P.; Visser, A. & Britton, C. 2007. RNA interference in parasitic nematodes of animals: a reality check. *Trends in Parasitology* 23:105-107.

Korayem, A.M.; Fabbri, M.; Takahashi, K.; Scherfer, C.; Lindgren, M.; Schmidt, O.; Ueda, R.; Dushay, M.S. & Theopold, U. 2004. A *Drosophila* salivary gland mucin is also expressed in immune tissues: evidence for a function in coagulation and the entrapment of bacteria. *Insect Biochemistry and Molecular Biology* 34(12):1297-1304.

Kurucz, E.; Markus, R.; Zsamboki, J.; Folki-Medzihradszky, K.; Darula, Z.; Vilmos, P.; Udvardy, A.; Krausz, I.; Lukacsovich, T.; Gateff, E.; Zettervall, C.J.; Hultmark, D. & Ando, I. 2007. Nimrod, a putative phagocytosis receptor with EGF repeats in *Drosophila* plasmatocytes. *Curr. Biol.* 17:649-654.

Kuwayama, H.; Yaginuma, T.; Yamashita, O. & Niimi, T. 2006. Germ-line transformation and RNAi of the ladybird beetle *Harmonia axyridis*. *Insect Mol. Biol.* 15:507-512.

Lamprou, I.; Mamali, I.; Dallas, K.; Fertakis, V.; Lampropoulou, M. & Marmaras, V.J. 2007. Distinct signalling pathways promote phagocytosis of bacteria, latex beads and lipopolysaccharide in medfly haemocytes. *Immunology* 121:314–327.

Lecellier, C.-H. & Voinnet, O. 2004. RNA silencing: no mercy for viruses? *Immunological Reviews* 198:285-303.

Lee, K.S.; Kim, B.Y.; Kim, H.J.; Seo, S.J.; Yoon, H.J.; Choi, Y.S.; Kim, I.; Han, Y.S.; Je, Y.H.; Lee, S.M.; Kim, D.H.; Sohn, H.D. & Jin, B.R. 2006. Transferrin inhibits stress-induced apoptosis in a beetle. *Free Radical Biology & Medicine* 41:1151–1161.

Lemaitre, B. & Hoffmann, J. 2007. The host defense of *Drosophila melanogaste*. *Annu. Rev. Immunol.* 25, 697-743

Levin, D.M.; Breuer, L.N.; Zhuang, S.F.; Anderson, S.A.; Nardi, J.B. & Kanost, M.R. 2005. A hemocyte-specific integrin required for hemocyte encapsulation in the tobacco hornworm, *Manduca sexta*. *Insect Biochemistry and Molecular Biology* 35:369-380.

Lynch, J.A. & Desplan, C. 2006. A method for parental RNA interference in the wasp *Nasonia vitripennis*. *Nat. Protoc.* 1:486-494.

Mao, C-P.; Lin, Y-Y.; Hung, C-F. & Wu, T-C. 2007. Immunological research using RNA interference technology. *Immunology* 121:295-307.

Marcus, J.M. 2005. Jumping genes and AFLP maps: transforming lepidopteran color pattern genetics. *Evolution & Development* 7(2):1008-114.

Martín, D.; Maestro, O.; Cruz, J.; Mane-Padros, D. & Belles, X. 2006. RNAi studies reveal a conserved role for RXR in molting in the cockroach *Blattella germanica*. *Journal of Insect Physiology* 52:410-416.

Meyering-Vos, M.; Merz, S.; Sertkol, M. & Hoffmann, K.H. 2006. Functional analysis of the allotostatin-A type gene in the cricket *Gryllus bimaculatus* and the armyworm *Spodoptera frugiperda*. *Insect Biochemistry and Molecular Biology* 36:492-504.

Napoli, C.; Lemieux, C. & Jorgensen, R. 1990. Introduction of a chimeric chalcone synthase gene into petunia results in reversible co-suppression of homologous genes in trans. *Plant Cell* 2:279-289.

Nishikawa, T. & Natori, S. 2001. Targeted disruption of a pupal hemocyte protein of *Sarcophaga* by RNA interference. *European Journal of Biochemistry* 268:5295-5299.

Novina, C.D. & Sharp, P.A. 2004. The RNAi revolution. *Nature* 430:161-164.

Obbard, D.J.; Jiggins, F.M.; Halligan, D.L. & Little, T.J. 2006. Natural selection drives extremely rapid evolution in antiviral RNAi genes. *Current Biology* 16(6):580-585.

Osta, M.A.; Christophides, G.K.; Vlachou, D. & Kafatos, F.C. 2004. Innate immunity in the malaria vector *Anopheles gambiae*: comparative and functional genomics. *Journal of Experimental Biology* 207(15):2551-2563.

Philips, J.A.; Rubin, E.J. & Perrimon, N. *Drosophila* RNAi screen reveals CD36 family member required for mycobacterial infection. *Science* 309(5738):1251-1253.

Quan, G.; Kanda, T. & Tamura, T. 2002. Induction of the *white egg 3* mutant phenotype by injection of the double-stranded RNA of the silkworm *white gene*. *Insect Molecular Biology* 11(3):217-222.

Rajagopal, R.; Sivakumar, S.; Agraval, N.; Malhotra, P. & Bhatnagar, R.K. 2002. Silencing of midgut aminopeptidase N of *Spodoptera litura* by double-stranded RNA establishes its role as *Bacillus thuringiensis* toxin receptor. *Journal of Biological Chemistry* 277(49):46849-46851.

Rämet, M.; Manfruelli, P.; Pearson, A.; Mathey-Prevot, B. & Ezekowitz, R.A.B. 2002. Functional genomic analysis of phagocytosis and identification of a *Drosophila* receptor for *E-coli*. *Nature* 416(6881):644-648.

Ratcliffe, N.A. 1993. Cellular defense responses of insects: unresolved problems. In: Beckage, N.E.; Thompson, S.N. & Federici, B.A. (Eds), *Parasites and Pathogens of Insects*, Academic Press, San Diego, CA, pp. 267–304.

Ratcliffe, N.A. & Whitten, M.M.A. 2004. Vector immunity. In: S.H. Gillespie; Smith, G.L. & Osbourn, A. (Eds), *Vector Immunity in Microbe–vector Interactions. SGM Symposium*, Cambridge University Press, Cambridge, pp. 199–262.

Reynolds, S.E & Eleftherianos, I. 2008. RNAi and insect immune system. In: Beckage, N.E. (Ed.), *Insect Immunology*, AP, San Diego, pp. 295-330.

Saleh, M.-C.; van Rij, R.P.; Hekele, A.; Gillis, A.; Foley, E.; O'Farrell, P.H. & Andino, R. 2006. The endocytic pathway mediates cell entry of dsRNA to induce RNAi silencing. *Nature Cell Biology* 8(8):793-U19

Scherfer, C.; Qazi, M.R.; Takahashi, K.; Ueda, R.; Dushay, M.S.; Theopold, U. & Lemaitre, B. 2006. The Toll immune-regulated *Drosophila* protein Fondue is involved in hemolymph clotting and puparium formation. *Developmental Biology* 295(1):156-163.

Shiao, S.-H.; Whitten, M.M.A.; Zachary, D.; Hoffmann, J.A. & Levashina, E.A. 2006. Fz2 and Cdc42 mediate melanization and actin polymerization but are dispensable for *Plasmodium* killing in the mosquito midgut. *PLOS Pathogens* 2(12):1152-1164.

Shin, S.W.; Kokoza, V.A. & Raikhel, A.S. 2003. Transgenesis and reverse genetics of mosquito innate immunity. *Journal of Experimental Biology* 206(21):3835-3843.

Strand, M.R. 2008. The insect cellular immune response. *Insect Science* 15:1-14.

Strand, M.R.; Beck, M.H.; Lavine, M.D. & Clark, K.D. 2006. *Microplitis demolitor* bracovirus inhibits phagocytosis by hemocytes from *Pseudoplusia includens*. *Archives of Insect Biochemistry and Physiology* 61(3):134-145.

Stroschein-Stevenson, S.L.; Foley, E.; O'Farrell, P.H. & Johnson, A.D. 2006. Identification of *Drosophila* gene products required for phagocytosis of *Candida albicans*. *PLOS Biology* 4(1):87-99.

Tanaka, H. & Suzuki, K. 2005. Expression profiling of a diapause-specific peptide (DSP) of the leaf beetle *Gastrophysa atrocyanea* and silencing of DSP by double-strand RNA. *Journal of Insect Physiology* 51:701–707.

Terenius, O.; Bettencourt, R.; Lee, S.Y.; Li, W.L.; Soderhall, K. & Faye, I. 2007. RNA interference of hemolin causes depletion of phenoloxidase activity in *Hyalophora cecropia*. *Dev Comp. Immunol*. 31:571-575

Theopold, U.; Schmidt, O.; Soderhall, K. & Dushay, M.S. 2004. Coagulation in arthropods: Defence, wound closure and healing. *Trends Immunol*. 25:289-294.

Tijsterman, M.; May, R.C.; Simmer, F.; Okihara, K.L. & Plasterk, R.H.A. 2004. Genes required for systemic RNA interference in *Caenorhabditis elegans*. *Curr. Biol*. 14:111-116.

Tomoyasu, Y.; Miller, S.C.; Tomita, S.; Schoppmeier, M.; Grossmann, D. & Bucher, G. 2008. Exploring systemic RNA interference in insects: a genome-wide survey for RNAi genes in *Tribolium*. *Genome Biology* 9(1):R10.1-R10.22.

Tsuzuki, S.; Sekiguchi, S.; Kamimura, M.; Kiuchi, M. & Hayakawa, Y. 2005. A cytokine secreted from the suboesophageal body is essential for morphogenesis of the insect head. *Mechanisms of Development* 122:189-197.

Turner, C.T.; Davy, M.W.; MacDiarmid, R.M.; Plummer, K.M.; Birch, N.P. & Newcomb, R.D. 2006. RNA interference in the light brown apple moth, *Epiphyas postvittana* (Walker) induced by double- stranded RNA feeding. *Insect Mol. Biol.* 15:383-391.

Valanne, S.; Kleino, A.; Myllymaki, H.; Vuoristo, J. & Ramet, M. 2007. Iap2 is required for a sustained response in the *Drosophila* Imd pathway. *Developmental and Comparative Immunology* 31(10):991-1001.

Vermehren, A.; Qazi, S. & Trimmer, B.A. 2001. The nicotinic alfa subunit MARA1 is necessary for cholinergic evoked calcium transients in *Manduca* neurons. *Neuroscience Letters* 313:113-116.

Vermehren, A. & Trimmer, B.A. 2005. Expression and function of two nicotinic subunits in insect neurons. *Journal of Neurobiology* 62:289-298.

Vilmos, P. & Kurucz, E. 1998. Insect immunity: evolutionary roots of the mammalian innate immune system. *Immunology Letters* 62:59-66.

Viney, M.E. & Thompson, F.J. 2008. Two hypotheses to explain why RNA interference does not work in animal parasitic nematodes. *International Journal for Parisotology* 38:43-47.

Wang, X.G.; Fuchs, J.F.; Infanger, L.C.; Rocheleau, T.A.; Hillyer, J.F.; Chen, C.C. & Christensen, B.M. 2005. Mosquito innate immunity: involvement of beta 1,3-glucan recognition protein in melanotic encapsulation immune responses in *Armigeres subalbatus*. *Molecular and Biochemical Parasitology* 139(1):65-73.

Wang, L.H. & Ligoxygakis, P. 2006. Pathogen recognition and signalling in the *Drosophila* innate immune response. *Immunobiology* 211(4):251-261.

Weinstock, G.M.; Robinson, G.E. & The Honeybee Genome Sequencing Consortium. 2006. Insights into social insects from the genome of the honeybee *Apis mellifera*. *Nature* 443:931-949.

Werner, T.; Borge-Renberg, K.; Mellroth, P.; Steiner, H. & Hultmark, D. 2003. Functional diversity of the *Drosophila* PGRP-LC gene cluster in the response to lipopolysaccharide and peptidoglycan. *Journal of Biological Chemistry* 278(29):26319-26322.

Winston, W.M.; Sutherlin, M.; Wright, A.J.; Feinberg, E.H. & Hunter, C.P. 2007. *Caenorhabditis elegans* SID-2 is required for environmental RNA interference. *Proc. Natl. Acad. Sci. USA* 104:10565-10570.

Zhao, M.; Soderhall, I.; Park, J.W.; Ma, Y.G.; Osaki, T.; Ha, N.-C.; Wu, C.F.; Soderhall, K. & Bok, L. 2005. A novel 43-kDa protein as a negative regulatory component of prophenoloxidase-induced melanin synthesis. *J. Biol. Chem.* 280:24744-24751.

In: Insect Viruses: Detection, Characterization and Roles
Editors: Ch. J. Connell and D. P. Ralston

ISBN: 978-1-60692-965-0
© 2009 Nova Science Publishers, Inc.

Chapter 3

Epidemiology of Kakugo Virus, An Insect Picorna-like Virus Identified in Aggressive Honeybee Workers – A Study of Virus-Host Interactions in the Honeybee Society

Tomoko Fujiyuki

Department of Biological Sciences, Graduate School of Science,
The University of Tokyo, 113-0033, Japan
Tel and Fax: +81-3-5841-4449
E-mail: fujiyuki_tomoko@yahoo.co.jp

Abstract

Honeybee (*Apis mellifera* L.) is a eusocial insect. Honeybees work together to create a nest (colony), obtain food, and maintain their shelter. In their colonies, female adult bees divide the various labors, such as reproduction and colony maintenance, to propagate effectively as a super-organism. Sociality and group living are not always beneficial, however, because animals in groups tend to attract predators. In addition, once an infectious disease occurs, it prevails in the group rather easily because of the dense population. Therefore, social animals may have specific defense systems for protection against predators and pathogens.

We have identified a novel insect picorna-like virus, Kakugo virus, from the brains of aggressive honeybees; that is, honeybees that counterattack their natural enemy, and studied its epidemiology. In this review, I discuss the interaction between Kakugo virus and the host honeybee, and briefly describe other insect picorna-like viruses.

Introduction

A honeybee colony comprises thousands of female adults, a few dozen of male bees (drones) from spring to autumn, pupae, larvae, and eggs. As is characteristic of eusocial insects, the female adults differentiate into two castes, a reproductive queen and sterile workers. The queen is engaged in reproduction, while the workers engage in all the other tasks required to maintain the colony. There is an age-related division of labor among the workers (age-polyethism) (Figure 1) (Winston, 1987). Young bees work inside the hive. Most of them nurse the queen and larvae by feeding them royal jelly, which is produced and secreted from the hypopharyngeal glands of the nurse bees (Kubo et al., 1996). In middle age, workers perform various labors, such as building the nest, cleaning, and guarding the hive (Breed, Robinson, and Page, 1990). Older workers go outside to collect pollen and nectar, and process it into honey. The sugar or protein-rich food stocked in the honeybee colony as well as larvae and pupae, provide an attractive target for predators.

The main predators are large mammals, such as bears and humans, and wasps such as hornets. Guard bees protect against natural enemies at the colony entrance by attacking and stinging the predators. The stinger, which has a hook-like shape, is connected to a venom sac. The stinger often detaches from the worker abdomen along with the venom sac, however, when the workers attack and sting their enemies, which allows for continuous injection of the venom into the predator (Dade, 1962). The loss of the stinger, however, is lethal to the bee, thus this attacking behavior is suicidal for the workers, providing advantages to the colony at the expense of the guard bee's own fitness. Such behavior by sterile individuals is called "altruistic behavior", a characteristic behavior of social animals (Wilson, 1975). The mechanism that regulates suicidal altruistic aggression of the workers is a biologically interesting subject and has been studied in the field of genetics and neurophysiology (Guzmán-Novoa et al., 2002; Guzman-Novoa and Page, 1994; Hunt, 2007; Hunt et al., 1998; Lobo et al., 2003; Uribe-Rubio et al., 2008).

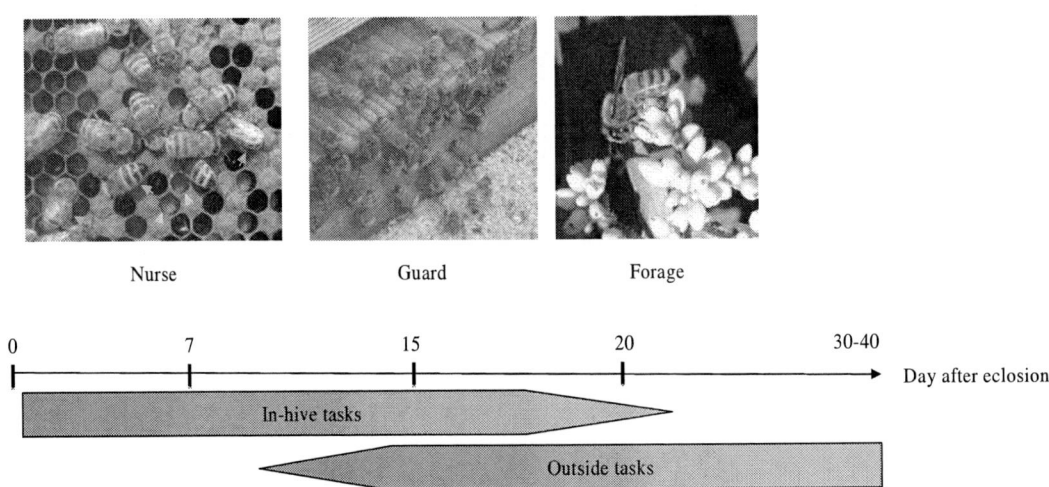

Figure 1. Age-related division of labor by worker bees. Nurse bees (arrows) take care of larvae in the comb (arrowheads).

In addition to its biologic importance, the honeybee is beneficial agriculturally and industrially. Bees pollinate agricultural crops and produce useful materials, such as honey, propolis, and royal jelly. In an effort to protect the honeybee from pathogenic microbe infections, microbial pathogens of the honeybee, such as bacteria, viruses, and fungi, have long been studied (Bailey, 1976; Bailey and Ball, 1991). To date, there are approximately 20 species of viruses known to infect the honeybees. This number is much greater than that known to infect other insect species, for example less than 10 species of viruses known to infect silkworm. Compared with lepidopteran baculoviruses, honeybee viruses are not well-documented which might indicate that a unique relationship has evolved between honeybees ad the viruses that infect them.

We examined the altruistic defense behavior of honeybees, which resulted in the identification of a novel virus. In this review, I discuss the relationship between the virus and the honeybee.

Kakugo Virus Identified from Aggressive Worker Bees

To understand the molecular and neural basis of the honeybee altruistic behaviors, we screened the candidate genes expressed preferentially in the brains of aggressive workers. As described in our previous paper and reviewed in detail (Fujiyuki et al., 2005), we used the giant hornet *Vespa mandarinia japonica,* a natural enemy of the honeybee, as a decoy to induce the guard aggression. After collecting the brains of aggressive workers that had attacked the giant hornet (attackers) and non-aggressive bees that had escaped from the hornet (escapers), we compared gene expression patterns in the brains using differential display (Figure 2). As a result, we identified a novel RNA from attacker brains. Because this novel RNA was found in attackers who seemed ready to attack and die, we named this RNA "Kakugo", which means the readiness in Japanese.

Surprisingly, *Kakugo* RNA (10152nt) is not transcribed from the honeybee genome, but is rather a genomic RNA of a novel insect picorna-like virus (Fujiyuki et al., 2004). It encodes a polyprotein homologous to those of various insect picorna-like viruses (Fig. 3). The material consist of *Kakugo* RNA shows the infectivity in the honeybee tissues and has a sedimentation coefficient similar to that of the virion particle of poliovirus. Furthermore, the sequence of *Kakugo* RNA is not found in the honeybee genome project (Honeybee genome sequencing consortium, 2006).

A number of RNA viruses have been identified and analyzed in insects (Gordon and Waterhouse, 2006). Among them, the genome sequences of 22 species of insect picorna-like viruses have been reported (Fig. 3). Insect picorna-like viruses comprise at least two subgroups, *Iflavirus* and *Cripavirus* (Chen and Siede, 2007; Fujiyuki et al., 2005). *Iflavirus* represents the Infectious flacherie virus (IFV) of *Bombyx mori*, the organism in which the genomic sequence was first determined (Isawa et al., 1998). IFV and Ifla-like viruses were assigned a genus, *Iflavirus*. *Cripavirus* represents Cricket paralysis virus (CrPV) and Cripa-like viruses. The major difference between the two groups is in the structure of the RNA genome. *Iflavirus* has a monocistronic RNA genome and *Cripavirus* has a dicistronic genome and is classified as *Dicistroviridae*. Thus, *Iflavirus* is more similar to mammalian picornaviruses, *Picornaviridae*, which has a monocistronic RNA genome. Thus, we determined that the novel *Iflavirus* (Kakugo virus: KV) had infected the brains of attackers.

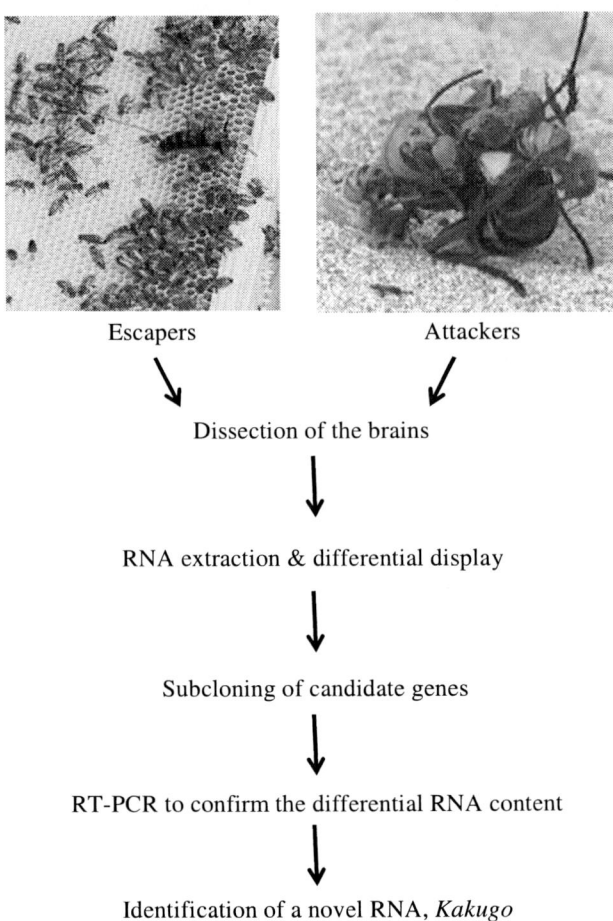

Figure 2. Process to identify *Kakugo* RNA from aggressive worker brains.
A giant hornet (arrowhead) was presented to workers, most inside workers escaped from the hornet (left, arrows), but guard bees attacked to it (right, arrows).

Prevalence of KV and Route of Infection in a Colony

An epidemiologic analysis of KV was performed to determine whether KV infects attackers selectively anytime or only under specific conditions. KV infection was examined in various workers, including attackers, escapers, foragers, nurse bees, and randomly collected bees derived from multiple colonies. As KV has an RNA genome, it can be measured by quantitative RT-PCR.

In the two colonies that we examined in our early study, KV was detected specifically in attackers, as previously reported. We also found colonies in which KV infected not only the attackers but other worker bees as well. The KV content was higher in those colonies than in the former colonies in which KV infected only the attackers. Thus, KV prevailed in such colonies. We followed the transmission of KV infection in one of the latter colonies, and the infection rate and KV content per bee increased over 10 days, suggesting that KV prevailed in the colony (Fujiyuki et al., 2006).

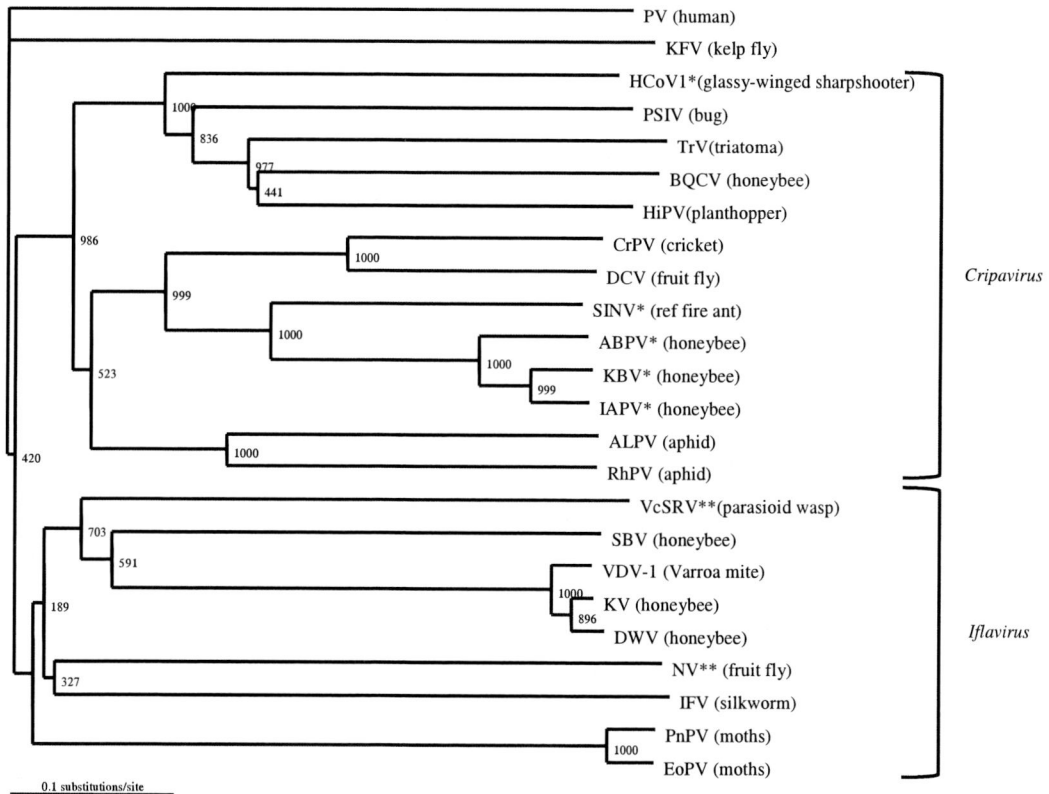

Figure 3. Phylogenetic tree of insect picorna-like viruses.
Phylogenetic tree constructed with the amino acid sequences encompassing the RdRp domain with the neighbor-joining method using the Clustal W program (Thompson, Higgins, and Gibsons, 1994). The virus species used here are KV (Fujiyuki et al., 2004), DWV (Lanzi et al., 2006), VDV-1 (Ongus et al., 2004), SBV (Ghosh et al., 1999), VcSRV (Reineke and Asgari, 2005), Nora virus (NV) (Habayeb, Ekengren, and Hultmark, 2006), IFV (Isawa et al., 1998), Perina nuda picorna-like virus (PnPV) (Wu et al., 2002), Ectropis obliqua picorna-like virus (Wang et al., 2004), Rhopalosiphum padi virus (RhPV) (Moon et al., 1998), aphid lethal paralysis virus (ALPV) (van Munster et al., 2002), IAPV (Maori et al., 2007), Kashmir bee virus (KBV) (DeMiranda et al., 2004), acute bee paralysis virus (ABPV) (Govan et al., 2000), *Solenopsis invicta* virus (SINV) (Valles et al., 2004), *Drosophila* C virus (DCV) (Johnson and Christian, 1998), CrPV (Wilson et al., 2000), Himetobi P virus (HiPV) (Nakashima, Sasaki, and Toriyama, 1999), Black queen cell virus (BQCV) (Leat et al., 2000), Triatoma virus (TrV) (Czibener et al., 2000), *Plautia stali* intestine virus (PSIV) (Sasaki et al., 1998), Homalodisca coagulata virus-1 (HoCV-1) (Hunnicutt et al., 2006), Kelp fly virus (KFV) (Hartley et al., 2005), and poliovirus (PV) as the outgroup (Nomoto et al., 1982). Numbers at the branches show bootstrap values obtained after 1000 replications of bootstrap sampling. The host animal of each virus is indicated in parentheses. Viruses marked with * or ** are not yet assigned as *Cripa-* or *Iflavirus* by the International Committee on Taxonomy of Viruses.

Furthermore, we observed that a parasitic mite, *Varroa destructor*, was parasitized in the KV-prevailed colony. The *Varroa* mite is a honeybee parasite that causes colony collapse (Bailey and Ball, 1991; Martin, 2001). The mite transmits pathogenic viruses to bees (Chen et al., 2004). KV was also found in the *Varroa* mite, and the cDNA sequences of the *Kakugo* RNA were identical with those detected from the worker bees (Figure 3). This strongly suggests that the *Varroa* mite transmits KV and causes KV prevalence in a colony.

These findings suggest that KV infects attackers selectively when the KV infection level in a colony is low, possibly at an early stage of infection. Infection exceeding a certain level might result in KV prevalence, possibly by increasing the infection rate and level in individual bees.

Phylogeny of Kakugo Virus and Close Relationship with Deformed Wing Virus

Different strains of the same virus species sometimes have different pathogenicity and infectivity. To examine whether KV comprises several different strains, we sequenced the partial cDNA of KV detected from various worker bees, and performed a phylogenetic analysis (Fujiyuki et al., 2006). The results indicated that multiple strains with slight mutations were present in one colony, and the strains were clustered by colony (Figure 4). A *Varroa* mite-infected colony contained several unrelated strains of KV, suggesting that the mites transmitted KV strains between colonies, in addition to within the colony. When we examined the relation between KV strains and the behaviors of KV infected workers, the KV strain detected in the attackers and the others was identical, indicating that workers can be infected with the same KV, independent of their task in a colony.

Next, I discuss the relationship of KV with other honeybee viruses. KV was first detected in Japan in 1999 and the sequence was published in 2004 (Fujiyuki et al., 2006; Fujiyuki et al., 2004). Afterward, it was revealed that KV is related to deformed wing virus (DWV) and Varroa destructor virus-1 VDV-1), which were identified and sequenced in Europe and the United States (Figures 3, 4).

DWV is a well-known virus that is believed to cause wing deformity of honeybee workers. The DWV infection is associated with the collapse of the honeybee colony and has been reported in many countries (Bailey and Ball, 1991). Lanzi and deMiranda determined the sequence of the DWV genome, revealing that DWV is 98% homologous to KV (Lanzi et al., 2006). VDV-1 was identified from the *Varroa* mite. Its genome has 89% homology with the KV and DWV genomes (Ongus et al., 2004).

Our phylogenetic analysis indicated that all of the KV strains are distinct from any of the DWV strains detected in Europe and the United States (Figure 4). In addition, analysis using DWV strains detected in various countries, including an Asian country (Nepal), showed that these strains are distinct from KV (Berényi et al., 2007). Therefore, KV, DWV, and VDV-1 are very closely related, but phylogenetically distinct, strains. Although it is controversial whether KV and DWV should be assigned as different viral species or rather strains of the same viral species, KV is at least a strain specific to Japan or East Asia, and may have distinct properties and pathogenicity.

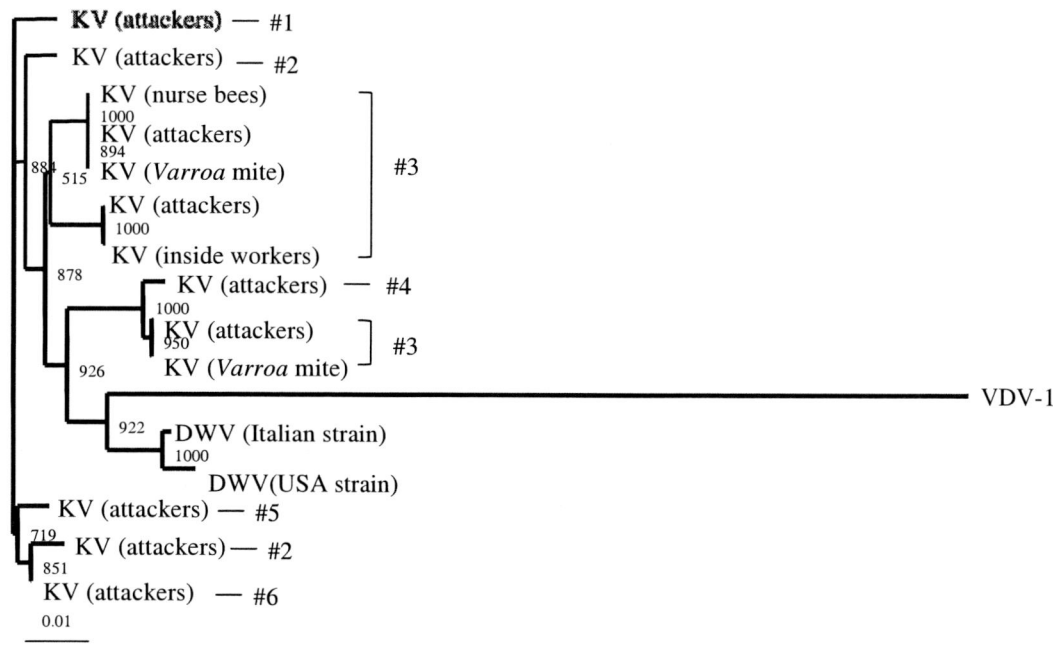

Figure 4. Phylogeny of KV strains and closely related viruses.
KV strains detected from workers and *Varroa* mites in three colonies are indicated. The behavior of the workers is shown in parentheses. Italian and US strains of DWV strains are included. Bar indicates distance.

Transmission by the *Varroa* mite is common to both DWV and KV. On the other hand, however, wing deformity associated with DWV infection has not yet been observed in KV-infected workers, including attackers. Rortais et al reported that there is no correlation between worker aggressive behavior and DWV infection (Rortais et al., 2006). This may be due to the different properties of DWV and KV. Given that under some condition KV infection is specific to attackers, it is possible that a more detailed epidemiologic analysis will reveal the conditions in which DWV and aggressive behavior correlate.

Recently, it was reported that DWV infection is associated with learning and memory deficits in worker bees (Iqbal and Mueller, 2007). A similar association of KV infection with this phenotype has not yet been evaluated. Because the induction of an immune response results in a learning deficit (Mallon, Brockmann, and Schmid-Hempel, 2003), the behavioral phenotype caused by DWV infection may be partly explained by the innate immune response.

To clarify whether the pathogenicity between KV and DWV is the same or differs, further analysis is needed, focusing on other inapparent infections that are common to bee viruses (Chen, Higgins, and Feldlaufer, 2005; Tentcheva et al., 2004; Yue and Genersch, 2005). Given that DWV was originally detected from bees collected in Japan (Bailey and Ball, 1991), the relation between KV and the Japanese strain of DWV also needs to be clarified. If the virulence of KV is in fact different from that of DWV, it will be necessary to determine how the 2% difference in the nucleotide sequence between the KV and DWV genomes results in a difference in the virulence.

Interactions between the Honeybee and the Viruses

Attackers infected with KV appear healthy and not physically different from normal bees. Further, KV was unknown until we found the genome in the course of the differential display analysis. Therefore, it is assumed that the coexistence between KV and the honeybee has evolved by the suppression of KV virulence, which may enable KV to remain in a bee or a colony for a long time. In addition, by establishing an avirulent relationship with the *Varroa* mite, KV may have acquired this route of transmission.

Other types of *Iflavirus* also suggest adaptive evolution in the hosts. *Venturia canescens* small RNA-containing virus (VcSRV) was identified from the endoparasitic wasp *V. canescens*, which parasitizes the host caterpillar *E. kuehniella* (Reineke and Asgari, 2005), but VcSRV has no obvious virulence to *V. canescens*. Interestingly, VcSRV was detected from parasitized *E. kuehniella* but not from unparasitized hosts, suggesting that the endoparasitic wasp injects the virus into their insect hosts during oviposition like polydnaviruses, which suppresses host immunity (Federici and Bigot, 2003; Summers and Dib-Hajj, 1995). Although the virulence of VcSRV in *V. canescens* or *E.kuehniella* is yet to be clarified, VcSRV may have a commensal relationship with *V. canescens*, which enables their larvae to develop in the parasitized host *E.kuehniella* and allows for an endosymbiotic relationship between the virus, the host, and the parasitoid's host.

Another *Iflavirus* is the infectious flacherie virus (IFV) of *Bombyx mori*, a pathogenic virus for silkworm that causes flacherie and death (Kawase, Hashimoto, and Nakagaki, 1980). IFV is not pathogenic to another host, however, the mulberry pyralid *Glyphodes pyloalis* (Watanabe et al., 1988). The silkworm disease, infectious flacherie, appears in the summer to fall rearing season in Japan and almost never continues into the next year in sericultural farms. Because IFV has no occlusion body, IFV is more easily inactivated under natural conditions than occluded baculovirus and cytoplasmic polyhedrosis virus. Thus, IFV infection in mulberry pyralid, which infests mulberry plantations, could contribute to transmission of the virus to the next season.

The low virulence of some insect picorna-like viruses may result not only from the rapid evolution of RNA viruses (Kinnunen et al., 1990; Martínez et al., 1991; Vignuzzi et al., 2006), but also from the long period during which the host insect co-evolved with the virus to suppress the virulence, thereby permitting co-existence, because insects are evolutionarily very ancient. This may have also contributed to viruses developing a transmission route. Host specificity of insect picorna-like viruses is not as restricted as mammalian picornaviruses, which have a rigid host range and are generally rarely transmitted by a vector animal (Masoumi, Hanzlik, and Christian, 2003; Plus et al., 1978).

The suicidal attack by bees infected with KV might have originally contributed to exclude the virus from the colony. Honeybees have various strategies to defend the colony from microbial pathogens. For example, workers show hygienic behaviors to find and throw away dead or diseased individuals from the colony (Rothenbuhler, 1964; Spivak and Gilliam, 1998a; Spivak and Gilliam, 1998b; Woodrow, 1942), and produce propolis, which has an antimicrobial role (Ghisalberti, 1979; Kujumgiev et al., 1999). Therefore, suicide attack may also contribute to eliminate pathogens.

Other picorna-like viruses are also known to influence honeybee worker behavior. Sacbrood virus (SBV) was identified from larvae that fail to pupate, become dark brown, and dry down to a flattened gondola-shaped scale (sacbrood) (Bailey, Gibbs, and Woods, 1964; Brčák, Svoboda, and Králík, 1963). SBV infection at the adult stage, however, does not cause any morphologic abnormality, but the lifespan of SBV-infected adults is shorter. Interestingly, SBV-infected workers become foragers earlier and fewer infected bees return to the colony (Bailey and Fernando, 1972). This has been interpreted as a strategy to remove SBV from the colony.

Very recently, the relation between Colony Collapse Disorder (CCD) and Israeli acute paralysis virus (IAPV) (Maori et al., 2007) was suggested. CCD is currently a big problem for apiculture, because most of the workers suddenly disappear from the colony, leaving the queen and brood behind (Oldroyd, 2007; Stokstad, 2007). Using metagenomic survey to find honeybee microbes that correlated with CCD, IAPV was identified as the most likely candidate pathogen (Cox-Foster et al., 2007). If IAPV-infected workers leave the colony, this may also be a behavioral strategy to defend the colony from the pathogen. An exception was also reported, however, that IAPV also infects non-CCD colonies. Perhaps another factor contributes to CCD together with IAPV infection.

Thus, it seems that the honeybee has acquired a defense response against microbes at the colony level by making use of behavioral changes of the pathogen-infected workers, which is helpful for excluding pathogens. On the other hand, the honeybee immune system is considered to have declined, as the honeybee has fewer immune-related genes than do the fruit fly, mosquito, and red flour beetle (Honeybee genome sequencing consortium, 2006; Evans et al., 2006; Kunieda et al., 2006; Zou et al., 2007). The decreased immune response at the individual level and the increased defense response at the colony level may have evolved in parallel, and therefore it may not have been necessary for the individual immune response of the honeybee to evolve to the same level as that in solitary insects, though it is still possible that a honeybee-specific molecular immunity mechanism had been acquired despite the smaller number of the genes.

It is documented that a relationship between host and parasite may influence animal behavior (Moore, 2002; Schmid-Hempel, 1998), therefore the evolution of the relationship between pathogens and the host honeybee may be related to their social behaviors.

Perspectives

Behavioral and physiologic alterations of animals due to virus-host interactions have become a recent focus in various scientific fields, such as immunology, neuroscience, and evolutionary ecology, as well as virology and medical science. In honeybee science, this possibility was proposed in the 1970s (Bailey and Fernando, 1972; Wang and Moeller, 1970), but progress during the last several decades has been slow. The development of molecular biologic technology has unexpectedly brought about the identification of novel viruses in studies with other aims (Ambrose and Clewley, 2006). In the case of insect picorna-like viruses, several viruses were identified from apparently healthy insects, by detecting the RNA genomes. The RNA genome of the nora virus of *Drosophila* was identified by differential display as was KV, and the genomes of Solenopsis invicta virus-1 of the red imported fire ant and Homalodisca coagulata virus-1 of the glassy-winged sharpshooter were found from expressed sequence tag analysis (Habayeb, Ekengren, and Hultmark, 2006; Hunnicutt et al., 2006; Valles et al., 2004). In addition, experimental techniques used to detect and quantify viral genomic RNA with higher sensitivity than traditional immunologic methods have been established (Blanchard et al., 2007; Celle et al., 2008; Chen, Higgins, and Feldlaufer, 2005; Grabensteiner et al., 2007; Kukielka et al., 2008).

On the other hand, it is important to recognize that viral infection of a sample with certain characteristics does not necessarily mean that the virus causes the symptoms (Kellam, 1998). The relationship between KV infection and honeybee aggression remains to be determined. We plan to perform the artificial infection experiments to examine the behavioral effects of KV, and continue epidemiologic analyses of the virus, which will enable us to clarify the causality between KV and honeybee behaviors and to better understand the KV-honeybee interaction. The following possibilities are especially important to examine: (i) KV infection may enhance worker aggression. (ii) Frequency of aggressive workers in a colony may correlate with the KV infection rate.

Purification of the material virus is a big barrier to performing artificial infection with insect picorna-like viruses. Among the approximately 20 honeybee viruses, 7 species/strains have been sequenced. All of them belong to insect picorna-like viruses (Fig. 3). The viruses have similar physical characteristics, which makes it difficult to isolate only one viral species using normal purification methods with ultra-centrifugation (Chen and Siede, 2007). To avoid contamination of viruses, infectious cDNA clones must be established. Many researchers have tried to obtain infectious cDNA clones of not only honeybee viruses but also other insect viruses. Only quite recently, however, have infectious transcripts have been attained for the Black queen cell virus of the honeybee (Benjeddou et al., 2002). One possible reason for the difficulty is that cloning techniques using *E.coli* and plasmids are not well-suited for the cloning step, unlike for mammalian picornaviruses. There are other hurdles with regard to honeybee viruses for which techniques in molecular biology have not been yet sufficiently progressed to the same level for other model insects such as *Drosophila* and *Bombyx mori*. For example, a honeybee-derived cell line has not yet been developed.

Very recently, an infectious cDNA clone of *Dicistroviridae* was obtained using baculovirus (Pal et al., 2007). Furthermore, completion of the genome project has promoted the molecular biology and virology studies in the honeybee. Thus, it is expected that an infectious clone of the honeybee will soon be obtained to enable artificial infection with a single virus.

Another approach to understand KV-honeybee interactions is to clarify the tissue tropism of KV in the honeybee. Because KV was identified from worker brains, we first hypothesized that only KV infection in the brain could affect neuronal activity, leading to behavioral alterations (Fujiyuki et al., 2004; Fujiyuki et al., 2005). Afterwards, it was revealed that KV also infects the thorax and abdomen (Fujiyuki et al., 2006). Thus, the possibility that physiologic changes in the whole body, such as the activation of immunity, could influence behaviors must be examined further (Adamo, 2002; Shedlowski, 2006). We are now examining the KV distribution in the brain and whole body in detail.

Conclusion

Intensive studies of the novel virus, KV, identified from aggressive worker brains are worthwhile, because KV infection appears to be related to the behavior of the host honeybee, in an intrinsic, rather than a pathogenic context. Many questions about KV still remain to be solved, such as the virologic characteristics of KV and the association of KV with the molecular basis of honeybee sociality. We hope to understand the virus-host interactions in the honeybee through the analysis of the KV- honeybee relationship.

Table 1. Non-pathogenic *Iflavirus* and the host animal

Iflavirus	Host	Another host
KV	Honeybee*	*Varroa* mite*
VcSRV	Endoparasitic wasp*	Mediterranean flour moth*
IFV	Silkworm	Mulberry pyralid*

* Pathogenicity is not found in the animal.

Table 2. Symptoms and behavioral influence of honeybee picorna-like viruses

Virus	Influence on physiology or development	Relation with adult behavior
Sacbrood virus	Malformation and death of larvae / Shortened lifespan of adult workers	Precocious aging / Decrease of the rate to return to colonies of foragers
Kakugo virus	No obvious virulence in adult workers	Specific infection to attackers under restricted condition
Deformed wing virus	Wing deformity	Deficit of learning and memory
Israeli acute paralysis virus	Paralysis of adult bees	Correlation with colony collapse disease

Acknowledgments

I would like to express my sincere gratitude to Prof. Takeo Kubo, Drs. Hideaki Takeuchi and Emiko Matsuzaka, Prof. Akio Nomoto, and Dr. Seii Ohka (The University of Tokyo),

Prof. Masato Ono, and Dr. Tetsuhiko Sasaki (Tamagawa University), Profs. Hans J Gross, Hildeburg Beier, Jürgen Tautz, Dr. Olaf Gimple, Ms. Klara Randolt (Würzburg Universität) for their collaborations. These studies were financially supported by the Program for Promotion of Basic Research Activities for Innovative Bioscience (PROBRAIN), Japan Society for the Promotion of Science, Naito Foundation, and Terumo Foundation. I also wish to express my sincere gratitude to Dr. Chikara Kaito for proofreading the draft of this chapter, and to Dr. Frank Columbus for giving me the opportunity to write this chapter.

References

Adamo, S. A. (2002). Modulating the modulators: Parasites, neuromodulators, and host behavioral change. Brain, Behavior, and Evolution 60, 370-377.

Ambrose, H. E., and Clewley, J. P. (2006). Virus discovery by sequece-independent genome amplification. Reviews in medical virology 16, 365-383.

Bailey, L. (1976). Viruses attacking the honey bee. Advances in Virus Research 20, 271-304.

Bailey, L., and Ball, B. V. (1991). "Honey bee pathology." Academic Press Inc., San Diego, CA.

Bailey, L., and Fernando, E. F. W. (1972). Effects of sacbrood virus on adult honey-bees. Annals of Applied Biology 72, 27-35.

Bailey, L., Gibbs, A. J., and Woods, R. D. (1964). Virology 23, 425-429.

Benjeddou, M., Leat, N., Allsopp, M., and Davidson, S. (2002). Development of infectious trascripts and genome manipulation of Black queen-cell virus of honey bees. Journal of General Virology 83, 3139-3146.

Berényi, O., Bakonyi, T., Derakhshifar, I., Köglberger, H., Topolska, G., Ritter, W., Pechhacker, H., and Nowotny, N. (2007). Phylogenetic analysis of deformed wing virus genotypes from diverse geographic origins indicates recent global distribution of the virus. Applied and Environmental Microbiology 73, 3605-3611.

Blanchard, P., Ribiere, M., Celle, O., Lallemand, P., Schurr, F., Olivier, V., Iscache, A. L., and Faucon, J. P. (2007). Evaluation of a real-time two-step RT-PCR assay for quantitation of Chronic bee paralysis virus (CBPV) genome in experimentally-infected bee tissues and in life stages of a symptomatic colony. Journal of Virological Methods 141, 7-13.

Brčák, J., Svoboda, J., and Králík, O. (1963). Electron microscopic investigation of sacbrood of honey bee. Journal of Insect Pathology 5, 385-386.

Breed, M. D., Robinson, G. E., and Page, J. R. E. (1990). Division of labor during honey bee colony defense. Behavioral Ecology and Sociobiology 27, 395-401.

Celle, O., Blanchard, P., Olivier, V., Schurr, F., Cougoule, N., Faucon, J. P., and Ribière, M. (2008). Detection of chronic bee paralysis virus (CBPV) genome and its replicative RNA form in various hosts and possible ways of spread. Virus Research 133, 280-284.

Chen, Y., Pettis, J. S., Evans, J. D., Kramer, M., and Feldlaufer, M. F. (2004). Transmission of kashmir bee virus by the ectoparasitic mite Varroa destructor. Apidologie 35, 441-448.

Chen, Y. P., Higgins, J. A., and Feldlaufer, M. F. (2005). Quantitative real-time reverse transcription-PCR analysis of deformed wing virus infection in the honeybee (Apis mellifera L.). Applied and Environmental Microbiology 71, 436-41.

Chen, Y. P., and Siede, R. (2007). Honey bee viruses. Advances in Virus Research 70, 33-80.

Cox-Foster, D. L., Conlan, S., Holmes, E. C., Palacios, G., Evans, J. D., Moran, N. A., Quan, P. L., Briese, T., Hornig, M., Geiser, D. M., Martinson, V., vanEngelsdorp, D., Kalkstein, A. L., Drysdale, A., Hui, J., Zhai, J., Cui, L., Hutchison, S. K., Simons, J. F., Egholm, M., Pettis, J. S., and Lipkin, W. I. (2007). A metagenomic survey of microbes in honey bee colony collapse disorder. Science 318, 283-7.

Czibener, C., La Torre, J. L., Muscio, O. A., Ugalde, R. A., and Scodeller, E. A. (2000). Nucleotide sequence analysis of Triatoma virus shows that it is a member of a novel group of insect RNA viruses. Journal of General Virology 81, 1149-1154.

Dade, H. A. (1962). "Anatomy and dissection of the honeybee." International Bee Research Association, Cardiff, UK.

DeMiranda, J. R., Drebot, M., Tyler, S., Shen, M., Cameron, C. E., Stoltz, D. B., and Camazine, S. M. (2004). Complete nucleotide sequence of Kasimir bee virus and comparison with acute bee paralysis virus. Journal of General Virology 85, 2263-2270.

Evans, J. D., Aronstein, K., Chen, Y. P., Hetru, C., Imler, J. L., Jiang, H., Kanost, M., Thompson, G. J., Zou, Z., and Hultmark, D. (2006). Immune pathways and defence mechanisms in honey bees Apis mellifera. Insect Molecular Biology 15, 645-656.

Federici, B. A., and Bigot, Y. (2003). Origin and evolucion of polydnaviruses by symbiogenesis of insect DNA viruses in endoparasitic wasps. Journal of Insect Pathology 49, 419-432.

Fujiyuki, T., Takeuchi, H., Ono, M., Ohka, S., Sasaki, T., Nomoto, A., and Kubo, T. (2004). Novel insect picorna-like virus identified in the brains of aggressive worker honeybees. Journal of Virology 78, 1093-1100.

Fujiyuki, T., Takeuchi, H., Ono, M., Ohka, S., Sasaki, T., Nomoto, A., and Kubo, T. (2005). Kakugo virus from brains of aggressive worker honeybees. Advances in Virus Research 65, 1-27.

Fujiyuki, T., Ohka, S., Takeuchi, H., Ono, M., Nomoto, A., and Kubo, T. (2006). Prevalence and phylogeny of Kakugo virus, a novel insect picorna-like virus that infects the honeybee (Apis mellifera L.), under various colony conditions. Journal of Virology 80, 11528-38.

Ghisalberti, E. (1979). Propolis: a review. Bee World 60, 59-84.

Ghosh, R. C., Ball, B. V., Willcocks, M. M., and Carter, M. J. (1999). The nucleotide sequence of sacbrood virus of the honey bee: an insect picorna-like virus. Journal of General Virology 80, 1541-1549.

Gordon, K. H. J., and Waterhouse, P. M. (2006). Small RNA viruses of insects: Expression in plants and RNA silencing. Advances in Virus Research 68, 459-502.

Govan, V. A., Leat, N., Allsopp, M., and Davison, S. (2000). Analysis of the complete genome sequence of acute bee paralysis virus shows that it belongs to the novel group of insect-infecting RNA viruses. Virology 277, 457-463.

Grabensteiner, E., Bakonyi, T., Ritter, W., Pechhacker, H., and Nowotny, N. (2007). Development of a multiplex RT-PCR for the simultaneous detection of three viruses of the honeybee (Apis mellifera L.): Acute bee paralysis virus, Black queen cell virus and Sacbrood virus. Journal of Invertebrate Pathology 94, 222-225.

Guzmán-Novoa, E., Hunt, G. J., Uribe, J. L., Smith, C., and Arechavaleta-Velasco, M. E. (2002). Confirmation of QTL effects and evidence of genetic dominance of honeybee defensive behavior: results of colony and individual behavioral assays. Behavior Genetics 32, 95-102.

Guzman-Novoa, E., and Page, J. R. E. (1994). Genetic dominance and worker interactions affect honebee colony defense. Behavioral Ecology 5, 91-97.

Habayeb, M. S., Ekengren, S. K., and Hultmark, D. (2006). Nora virus, a persistent virus in Drosophila, defines a new picorna-like virus family. Journal of General Virology 87, 3045-3051.

Hartley, C. J., Greenwood, D. R., Gilbert, R. J. C., Masoumi, A., Gordon, K. H. J., Hanzlik, T. N., Fry, E. E., Stuart, D. I., and Scotti, P. D. (2005). Kelp fly virus: a novel group of insect picorna-like viruses as defined by genome sequence analysis and a distinctive virion structure. Journal of Virology 79, 13385-13398.

Honeybee genome sequencing consortium. (2006). Insights into social insects from the genome of the honeybee Apis mellifera. Nature 443, 931-949.

Hunnicutt, L. E., Hunter, W. B., Cave, R. D., Powell, C. A., and Mozoruk, J. J. (2006). Genome sequence and molecular characterization of Homolodisca coagulata virus-1, a novel virus discovered in the glassy-winged sharpshooter (Hemiptera: Cicadellidae). Virology 6350, 67-78.

Hunt, G. J. (2007). Flight and fight: a comparative view of the neurophysiology and genetics of honey bee defensive behavior. Journal of Insect Physiology 53, 399-410.

Hunt, G. J., Guzman-Novoa, E., Fondrk, M. K., and Page, J. R. E. (1998). Quantitative trait loci for honey bee stinging behavior and body size. Genetics 148, 1203-1213.

Iqbal, J., and Mueller, U. (2007). Virus infection causes specific learning deficits in honeybee foragers. Proceedings of Biological Sciences 274, 1517-1521.

Isawa, H., Asano, S., Sahara, K., Iizuka, T., and Bando, H. (1998). Analysis of genetic information of an insect picorna-like virus, infectious flacherie virus of silkworm: evidence for evolutionary relationships among insect, mammalian and plant picorna(-like) viruses. Archives of Virology 143, 127-143.

Johnson, K. N., and Christian, P. D. (1998). The novel genome organization of the insect picorna-like virus Drosophila C virus suggests this virus belongs to a previously undescribed virus family. Journal of General Virology 79, 191-203.

Kawase, S., Hashimoto, Y., and Nakagaki, M. (1980). Characterization of flacherie virus of the silkworm, Bombyx mori. Journal of Sericultural Science of Japan 49, 477-484.

Kellam, P. (1998). Molecular identification of novel viruses. Trends in Microbiology 6, 160-166.

Kinnunen, L., Huovilainen, A., Pöyry T., and Hovi, T. (1990). Rapid molecular evolution of wild type 3 poliovirus during infection in individual hosts. Journal of General Virology 71, 317-324.

Kubo, T., Sasaki, M., Nakamura, J., Sasagawa, H., Ohashi, K., Takeuchi, H., and Natori, S. (1996). Change in the expression of hypopharyngeal-gland proteins of the worker honeybees (Apis mellifera L.) with age and/or role. Journal of Biochemistry 119, 291-295.

Kujumgiev, A., Tsvetkova, I., Serkedjieva, Y., Bankova, V., Christov, R., and Popov, S. (1999). Antibacterial, antifungal and antiviral activity of propolis of different geographic origin. Journal of Ethnopharmacology 64, 235-240.

Kukielka, D., Esperón, F., Higes, M., and Sánchez-Vizcaíno, J. M. (2008). A sensitive one-step real-time RT-PCR method for detection of deformed wing virus and black queen cell virus in honeybee Apis mellifera. Journal of Virological Methods 147, 275-281.

Kunieda, T., Fujiyuki, T., Kucharski, R., Foret, S., Ament, S. A., Toth, A. L., Ohashi, K., Takeuchi, H., Kamikouchi, A., Kage, E., Morioka, M., Beye, M., Kubo, T., Robinson, G. E., and Maleszka, R. (2006). Carbohydrate metabolism genes and pathways in insects: insights from the honey bee genome. Insect Molecular Biology 15, 563-76.

Lanzi, G., de Miranda, J. R., Boniotti, M. B., Cameron, C. E., Lavazza, A., Capucci, L., Camazine, S. M., and Rossi, C. (2006). Molecular and biological characterization of deformed wing virus of honeybees (Apis mellifera L.). Journal of Virology 80, 4998-5009.

Leat, N., Ball, B., Govan, V., and Davidson, S. (2000). Analysis of the complete genome sequence of black queen-cell virus, a picorna-like virus of honey bees. Journal of General Virology 81, 2111-2119.

Lobo, N. F., Ton, L. Q., Hill, C. A., Emore, C., Romero-Severson, J., Hunt, G. J., and Collinf, F. H. (2003). Genomic analysis in the sting-2 quantitative trait locus for defensive behavior in the honey bee, Apis mellifera. Genome Research 13, 2588-2593.

Mallon, E. B., Brockmann, A., and Schmid-Hempel, P. (2003). Immune response inhibits associative learning in insects. . Proceedings of the Royal Society of London. Series B, Biological sciences. 270, 2471-2473.

Maori, E., Lavi, S., Mozes-Koch, R., Gantman, Y., Peretz, Y., Edelbaum, O., Tanne, E., and Sela, I. (2007). Isolation and characterization of Israeli acute paralysis virus, a dicistrovirus affecting honeybees in Israel: evidence for diversity due to intra- and inter-species recombination. Journal of General Virology 88, 3428-3438.

Martin, S. J. (2001). The role of Varroa and viral pathogens in the collapse of honeybee colonies: a modeling approach. Journal of Applied Ecology 38, 1082-1093.

Martínez, M. A., Carrillo, C., González-Candelas, F., Domingo, M. E., and Sobrino, F. (1991). Fitness alteration of foot-and-mouth disease virus mutants: Measurement of adaptability of viral quasispecies. Journal of Virology 65, 3954-3957.

Masoumi, A., Hanzlik, T. N., and Christian, P. D. (2003). Functionality of the 5'- and intergenic IRES elements of cricket paralysis virus in a range of insect cell lines, and its relationship with viral activities. Virus Research 94, 113-120.

Moon, J. S., Domier, L. L., McCoppin, N. K., D'Arcy, C. J., and Jin, H. (1998). Nucleotide sequence analysis shows that Rhopalosiphum padi virus is a member of a novel group of insect-infecting RNA viruses. Virology 243, 54-65.

Moore, J. (2002). "Parasites and the behavior of animals." Oxford University press, New York, NY.

Nakashima, N., Sasaki, J., and Toriyama, S. (1999). Determining the nucleotide sequence and capsid-coding region of Himetobi P virus: a member of a novel group of RNA viruses that infect insects. Archives of Virology 144, 2051-2058.

Nomoto, A., Omata, T., Toyoda, H., Kuge, S., Horie, H., Kataoka, Y., Genba, Y., Nakano, Y., and Imura, N. (1982). Complete nucleotide sequence of the attenuated poliovirus Sabin 1 strain genome. Proceedings of National Academy of Sciences of the United States of America 79, 5793-5797.

Oldroyd, B. P. (2007). PLoS Biology e168, 5.

Ongus, J. R., Peters, D., Bonmatin, J. M., Bengsch, E., Vlak, J. M., and van Oers, M. M. (2004). Complete sequence of a picorna-like virus of the genus Iflavirus replicating in the mite Varroa destructor. Journal of General Virology 85, 3747-3755.

Pal, N., Boyapalle, S., Beckett, R., Miller, W. A., and Bonning, B. C. (2007). A baculovirus-expressed dicistrovirus that is infectious to aphids. Journal of Virology 81, 9339-9345.

Plus, N., Croizier, G., Reinganum, C., and Scotti, P. D. (1978). Cricket paralysis virus and Drosophila C virus: erological analysis and comparison of capsid polypeptides and host range. Journal of Invertebrate Pathology 31, 296-302.

Reineke, A., and Asgari, S. (2005). Presence of a novel small RNA-containing virus in a laboratory culture of the endoparasitic wasp Venturia canescens (Hymenoptera: Ichneumonidae). Journal of Insect Physiology 51, 127-135.

Rortais, A., Tentcheva, D., Papachristoforou, A., Gauthier, L., Arnold, G., Colin, M. E., and Bergoin, M. (2006). Deformed wing virus is not related to honey bees' aggressiveness. Virology Journal 3, 61.

Rothenbuhler, W. C. (1964). Behaviour genetics of nest cleaning in honey bees. IV. Response of F1 and backcross generations to disease-killed brood. American Zoologist 4, 111-123.

Sasaki, J., Nakashima, N., Saito, H., and Noda, H. (1998). An insect picorna-like virus, Plautia stali intestine virus, has genes of capsid proteins in the 3' part of the genome. Virology 244, 50-58.

Schmid-Hempel, P. (1998). "Parasites in social insects." Princeton University Press, Chichester, West Sussex.

Shedlowski, M. (2006). Insecta immune-cognitive interactions. Brain, Behavior, and Immunity 20, 133-134.

Spivak, M., and Gilliam, M. (1998a). Hygienic behaviour of honey bees and its application for control of brood diseases and varroa. Part I. Hygienic behaviour and resistance to American foulbrood. Bee World 79, 124-134.

Spivak, M., and Gilliam, M. (1998b). Hygienic behaviour of honey bees and its application for control of brood diseases and varroa. Part II. Studies on hygienic behaviour since the Rothenbuhler era. Bee World 79, 169-186.

Stokstad, E. (2007). The case of the empty hives. Science 316, 970.

Summers, M. D., and Dib-Hajj, S. D. (1995). Polydnavirus-facilitated endoparasite protection against host immune defenses. Proceedings of National Academy of Sciences of the United States of America 92, 29-36.

Tentcheva, D., Gauthier, L., Zappulla, N., Dainat, B., Cousserans, F., Colin, M. E., and Bergoin, M. (2004). Prevalence and seasonal variations of six bee viruses in Apis mellifera L. and Varroa destructor mite populations in France. Applied and Environmental Microbiology 70, 7185-7191.

Thompson, J. D., Higgins, D. G., and Gibsons, T. J. (1994). CLUSTAL W: improving the sensitivity of progressive multiple sequence alignment through sequence weighting, position-specific gap penalties and weight matrix choice. Nucleic Acids Research 22, 4673-4680.

Uribe-Rubio, J. L., Guzmán-Novoa, E., Vázquez-Peláez, C. G., and Hunt, G. J. (2008). Genotype, task specialization, and nest environment influence the stinging response thresholds of individual Africanized and European honeybees to electrical stimulation. Behavior Genetics 38, 93-100.

Valles, S. M., Strong, C. A., Dang, P. M., Hunter, W. B., Pereira, P. M., Oi, D. H., Shapiro, A. M., and Williams, D. F. (2004). A picorna-like virus from the red imported fire ant, Solenopsis invicta: initial discovery, genome sequence, and characterization. Virology 328, 151-157.

van Munster, M., Dullemans, A. M., Verbeek, M., van den Heuvel, J. F. J. M., Clérivet, A., and van der Wilk, F. (2002). Sequence analysis and genomic organization of Aphid lethal paralysis virus: a new member of the family Dicistroviridae. Journal of General Virology 83, 3131-3138.

Vignuzzi, M., Stone, J. K., Arnold, J. J., Cameron, C. E., and Andino, R. (2006). Quasispecies diversity determines pathogenesis through cooperative interactions within a viral population. Nature 439, 344-348.

Wang, D.-I., and Moeller, F. E. (1970). The division of labor and queen attendance behavior of Nosema-infected worker honey bees. Journal of Economic Entomology 63, 1539-1541.

Wang, X., Zhang, J., Lu, J., Yi, F., Liu, C., and Hu, Y. (2004). Sequence analysis and genomic prganization of a new insect picorna-like virus, Ectropis obliqua picorna-like virus, isolated from Ectropis obliqua. Journal of General Virology 85, 1145-1151.

Watanabe, H., Kurihara, Y., Wang, Y.-X., and Shimizu, T. (1988). Mulberry pyralid, Glyphodes pyloalis: habitual host of nonocculuded viruses pathogenic to the silkworm, Bombyx mori. Journal of Invertebrate Pathology 52, 401-408.

Wilson, E. O. (1975). "Sociobiology: the new synthesis." The Belknap Press of Harvard University Press.

Wilson, J. E., Powell, M. J., Hoover, S. E., and Sarnow, P. (2000). Naturally occurring dicistronic cricket paralysis virus RNA is regulated by two internal ribosome entry sites. Molecular and Cellular Biology 20, 4990-4999.

Winston, M. L. (1987). "The biology of the honeybee." Harvard University Press, Cambridge, MA.

Woodrow, A. W. (1942). The mechanism of colony resistance to American foulbrood. Journal of Economic Entomology 35, 327-330.

Wu, C. Y., Lo, C. F., Huang, C. J., Yu, H. T., and Wang, C. H. (2002). The complete genome sequence of Perina nuda picorna-like virus, an insect-infecting RNA virus with a genome organization similar to that of the mammalian picornaviruses. Virology 294, 312-323.

Yue, C., and Genersch, E. (2005). RT-PCR analysis of Deformed wing virus in honeybees (Apis mellifera) and mites (Varroa destructor). Journal of General Virology 86, 3419-24.

Zou, Z., Evans, J. D., Lu, Z., Zhao, P., Williams, M., Sumathipala, N., Hetru, C., Hultmark, D., and Jiang, H. (2007). Comparative genomic analysis of the Tribolium immune system. Genome Biology 8, R177.

Chapter 4

Honeybee Viruses in Uruguay: First Detection of Honeybee Viruses in South America and Their Potential Role in Mortality of Honeybees

Karina Antúnez, Bruno D'Alessandro and Pablo Zunino
Departamento de Microbiología, Instituto de Investigaciones Biológicas Clemente Estable, Avenida Italia 3318, CP 11600, Montevideo, Uruguay

Abstract

More than 18 RNA viruses that affect honeybees have been reported. Most of them cause unapparent infections or persist in a latent form in healthy colonies, while only a few viruses produce clinical symptoms easily recognizable by beekeepers. The presence of these viruses has been associated with honeybee mortality episodes that have been taking place during the last years worldwide. Several reports have been recently published in North America and Europe, although precise information about bees' mortality and the presence of viruses in South America was lacking.

In the present work, we report the presence of different RNA viruses, including acute bee paralysis virus, chronic bee paralysis virus, black queen cell virus, sacbrood virus and deformed wing virus in honeybee samples from different locations in Uruguay. This was the first report about the presence of potentially pathogenic honeybee viruses in South America, and their relation with other honeybee pathogens such as *Varroa destructor* and *Nosema* spp. is discussed.

The detection of viruses in different geographic regions in the country, the simultaneous co-infection of colonies by several viruses and even other pathogens, and the fact that in most of the samples one or more viruses were found, indicate that they are widely spread in the region.

Introduction

The honeybee *Apis mellifera* produces a wide variety of products that have been used by humans for centuries with industrial, medicinal and alimentary objectives. Honey, pollen, propolis, royal jelly and beeswax can be considered the most important products derived from the activity of honeybees. In addition, honeybees play an essential role in the ecology of natural environments through pollination. It ensures fecundation and fructification of different vegetable species, conservation of species in danger, biological diversity and sustentation of plants that control erosion, being essential for the conservation of natural ecosystems and increase of agricultural production. It has been estimated that one third of the human diet can be traced to bee pollination. The value of honey bee pollination for agriculture has been estimated at more than $14.6 billion in the United States and $443 million in Canada [Delaplane and Mayer, 2000; Morse and Calderone, 2000]. According to FAO and the European Union, the value of pollination is 20 to 30 times higher than the value of honey production.

However, due to their communal and social life style, honeybees are susceptible to a great variety of pathogens [Morse and Flottum, 1997; Schmid-Hempel, 1998]. Among these pathogens, viruses are one of the major threats to the health of honeybees.

More than eighteen viruses that affect honeybees have been described, including acute bee paralysis virus (ABPV), chronic bee paralysis virus (CBPV), black queen cell virus (BQCV), sacbrood virus (SBV), deformed wing virus (DWV) and Kashmir virus (KV) [Chen and Siede, 2007].

CBPV was one of the first isolated viruses that affect adult honeybees, causing a disease characterized by bee paralysis, trembling, flightlessness and, sometimes, black individuals crawling at the hive entrance (Bailey et al., 1963, Bailey, 1975). ABPV also affects adult honeybees and causes similar symptoms, but it is more virulent than CBPV [Bailey et al., 1963]. BQCV was first detected in queen larvae and prepupae that became brown to black [Bailey and Woods, 1974]. However, it also affects larvae and pupae from worker bees without developing typical symptoms. SBV affects larvae of honeybees that acquire a pale yellow colour; their skin becomes leathery and the ecsydial fluid accumulates between the body and the skin. It also affects adult bees but without developing typical symptoms [Ball and Bailey, 1997]. DWV affects adult bees causing a well-defined disease, characterized by deformed, crumpled or shrunken wings and decrease of body size. However, it has also been detected in eggs, larvae and pupae [Ball and Bailey, 1997]. KV affects all stages of the bee life cycle, causing mortality, but it does not cause specific symptoms.

Although viruses may be associated with different symptoms, all of them can persist in a latent state in apparently healthy colonies [Allen and Ball, 1996; Bailey, 1967].

The genomic sequences of the different viruses have been determined, and the genetic characteristics have been described (Table 1) [Ghosh et al., 1999; Govan et al., 2000; Leat et al., 2000; De Miranda et al., 2004; Lanzi et al., 2006].

Table 1. Taxonomy and genomic characteristics of different honeybee viruses

Name	Genus	Family	Genbank accesion n°	Reference	Genome organziation
ABPV	Cripavirus	Dicistroviridae	NC_002548	Govan et al., 2000	monopartite bicistronic
BQCV	Cripavirus	Dicistroviridae	NC_3784	Leat el al., 2000	monopartite bicistronic
DWV	Iflavirus	na	NC_004830	Lanzi et al., 2006	monopartite monocistronic
SBV	Iflavirus	na	NC_002066	Ghosh et al., 1999	monopartite monocistronic
KV	Cripavirus	Dicistroviridae	NC_004807	De Miranda et al., 2004	monopartite bicistronic

na: not assigned.

The importance of viruses in honeybee health has gained attention in recent years, especially due to the serious problem of honeybee mortality occurring worldwide. Several factors may play a role in this process, the presence of RNA viruses being one of the potential causes, along with the presence of other pathogens such as *Varroa destructor, Nosema ceranae* or even exposure to insecticides or pesticides used in agriculture [Suchail et al., 2004; Higes et al., 2006].

In Uruguay, a country with a population of about three million people, there are about 4,000 beekeepers and more than 400,000 beehives, and in recent years honey has become one of the most important agricultural products for export. In previous work, we reported the presence of ABPV, CBPV, BQCV, SBV and DWV in Uruguayan honeybees by RT-PCR. These studies were the first reports of viruses in bees from South America [Antúnez et al., 2005; Antúnez et al., 2006]. Due to the importance of honeybees for the sustainability of natural environments and agriculture production in Uruguay and worldwide, and the great losses associated with bee mortality suffered every year, it is greatly important to elucidate these phenomena.

The aim of the present chapter is to report the presence of different viruses (ABPV, CBPV, BQCV, SBV, KV and DWV) in Uruguayan honeybees, and to discuss their relationship with the incidence of other bee pathogens such as *V. destructor* and *Nosema* spp. and the development of mortality-associated symptoms.

Materials and Methods

Samples

Eighty-eight honeybee samples (*A. mellifera*) from different provinces of Uruguay (Colonia and Soriano [west], Canelones and San José [south], Maldonado, Treinta y Tres and Lavalleja [east] and Rivera [north]) were used in this study.

Samples were collected during two years, from December 2003 to December 2005. Sub-samples were sent refrigerated to the Laboratory of Microbiology, IIBCE (Montevideo, Uruguay) for the analysis of viruses, and to the Laboratory of Apiculture, INIA (Colonia, Uruguay) for the analysis of *Nosema* spp. and *V. destructor*.

RNA Extraction

Ten honeybees were randomly selected from each sample and placed in sterile plastic bags using sterile forceps and 10 ml of sterile phosphate saline buffer (PBS) was added. Bees were crushed for 2 min at high speed in a Stomacher 80 Lab Blender (Seward, London, UK) and the resultant homogenate was first centrifuged at 1500 x g for 10 min. The supernatant was recovered and centrifuged again at 12000 x g for 15 min and 140 µl of the final supernatant was used for viral RNA extraction.

RNA was extracted using a QIAamp Viral RNA Mini Kit (Qiagen, Hilden, Germany) according to the manufacturer's instructions.

RT-PCR

Reverse transcription of RNA and amplification of cDNA were performed using a continuous RT-PCR method (One Step RT-PCR kit, Qiagen), according to the manufacturer's recommendations and conditions. Primers used in the reactions are described in Table 2. Negative RT-PCR controls were carried out excluding nucleic acids from the reaction and adding RNA from healthy honeybees. The RT-PCR program included a reverse transcription stage at 50°C for 30 min, followed by an initial PCR activation step at 95°C for 15 min. This was followed by 40 cycles of 94°C for 1 min, 55°C for 1 min and 72°C for 1 min, and a final extension step at 72°C for 10 min. The reaction was performed using a T1 Biometra Thermocycler and products were visualized by electrophoresis in 0.8% (w/v) agarose gels stained with ethidium bromide [Sambrook et al., 1989].

Data Analysis

Data analysis was performed using the VennMaster tool for drawing Venn/Euler-diagrams, an open source Java application, which is available online [Kestler et al., 2008]. The data were arranged categorically in a tab delimited file format, placing a sample/category pair in each line, and this was the source for the program VennMaster. The sizes of the circular areas are proportional to the amount of category members.

Table 2. Primers used for the detection of CBPV, ABPV, BQCV, SBV, DWV and KV

Primer	Sequence (5'-3')	Length (bp)	Amplification target	Reference
ABPV1	TTATGTGTCCAGAGACTGTATCCA	900	RNA polymerase gene	Benjeddou et al., 2001
ABPV2	GCTCCTATTGCTCGGTTTTTCGGT			Benjeddou et al., 2001
CBPV1	AGTTGTCATGGTAACAGGATACGAG	455	viral capsid gene	Ribiere et al., 2002

Table 2. (Continued)

Primer	Sequence (5'-3')	Length (bp)	Amplification target	Reference
CBPV2	TCTAATCTTAGCACGAAAGCCGAG			Ribiere et al., 2002
BQCV1	TGGTCAGCTCCCACTACCTTAAACI	700	structural polyprotein gene	Benjeddou et al., 2001
BQCV2	GCAACAAGAAGAAACGTAAACCACI			Benjeddou et al., 2001
SBV1	GGATGAAAGGAAATTACCAG	426	polyprotein gene	Tentcheva et al., 2004b
SBV2	CCACTAGGTGATCCACACT			Tentcheva et al., 2004b
DWV1	TTTGCAAGATGCTGTATGTGG	395	gene for polyprotein	Tentcheva et al., 2004a.
DWV2	GTCGTGCAGCTCGATAGGAT			Tentcheva et al., 2004a.
KV1	GATGAACGTCGACCTATTGAA	393	RNA polymerase gene	Stoltz et al., 1995
KV2	TGTGGGTTGGCTATGAGTTCA			Stoltz et al., 1995

Results

Survey 2004

In a first survey (from December 2003 to December 2004) we analyzed 36 samples of worker honey bees from different provinces of Uruguay. Then we focused on the detection of two of the most studied viruses at that time, ABPV and CBPV, and our results showed that 78% of the samples were infected with at least one virus (Figure 1A).

A high proportion of the infected samples showed symptoms of disease; specifically 50% of the samples were associated to mortality, while 32% were apparently healthy (Figure 1B). However, disease symptoms were also present in non infected samples, or more precisely, in samples in which ABPV and CBPV were not detected since the presence of other viruses cannot be discarded.

Samples from almost all the provinces showed the presence of virus and we found also a great proportion of co-infected samples with both ABPV and CBPV (43%) (Figure 1C). The presence of mortality symptoms was not associated to one particular virus, or even with co-infection, since it was distributed almost equal between ABPV and CBPV. Also within co-infected samples the presence or absence of symptoms were distributed almost equally.

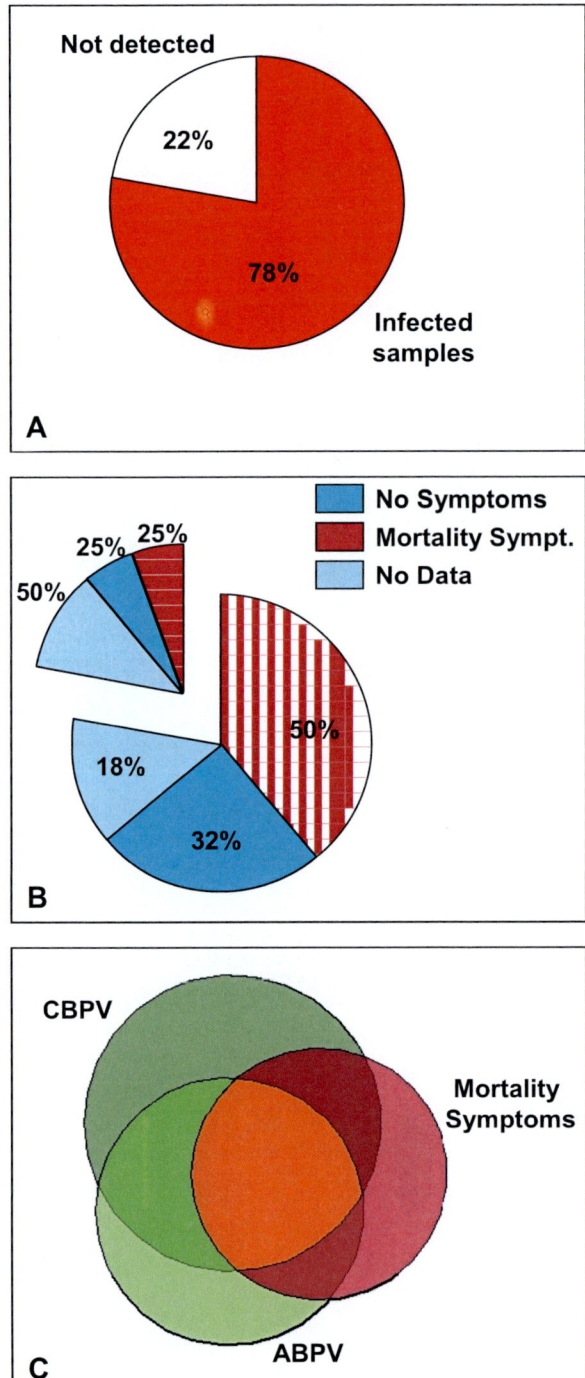

Figure 1. Analysis of the presence of CBPV and ABPV in honeybees from Uruguay in a survey carried out during 2005.
A. Proportion of viral infection in the samples.
B. Distribution of observed symptoms.
C. Relations between the presence of different viruses and mortality symptoms.

Survey 2005

In the second stage of our study, a survey that took place from May 2005 to December 2005, we analyzed 52 samples of worker honey bees again from different provinces of Uruguay. In this survey we included the detection of a broad range of viruses, in an attempt to explain the proportion of non infected samples with symptoms associated to mortality previously observed. For the analysis of the data, the survey was divided in winter and summer samples.

Winter Samples

In these samples detection of BQCV was added to ABPV and CBPV.

The analysis of the results shows that almost all the winter samples were infected with at least one virus (94 %). A high proportion of the infected colonies were asymptomatic and the only two samples that were negative for all three viruses showed mortality symptoms (Figure 2A-B).

The first highlight of the results compared to the previous year, is the low proportion of ABPV detection (6%), and the absence of co-infection between ABPV and CBPV. Nevertheless, the proportion of CBPV-infected samples during this season (72%) was almost the same than the proportion observed in the previous year (75%). The second highlight is the high incidence of BQCV in the colonies (about 88%), and as a consequence 77% of the infected samples were co-infected either by BQCV and CBPV (70 %) or BQCV and ABPV (7%).

The presence of mortality was not associated with one virus in particular, although half of the colonies that showed mortality-associated symptoms were co-infected by BQCV and CBPV or BQCV and ABPV (Figure 2C).

Summer Samples

In these samples detection of SBV, DWV and KV was added to ABPV, CBPV and BQCV.

In the summer samples the percentage of viral infection increased to 100%, being all of them co-infected by BQCV, SBV and DWV which suggests that these viruses are established in the colonies of Uruguay (Figure 3A). However, mortality-associated symptoms were detected only in 30% of the samples and all these samples were infected by these three viruses and one was even additionally co-infected with CBPV (Figure 3B-C).

The proportion of CBPV-infected samples was much lower than in winter and the absence of co-infected samples with CBPV and ABPV persisted. When the results of the previous year were analyzed, a reduction in CBPV detection was also registered from winter to summer. However the amount of samples in summer was much lower than in winter, so further studies are required to propose an accurate conclusion.

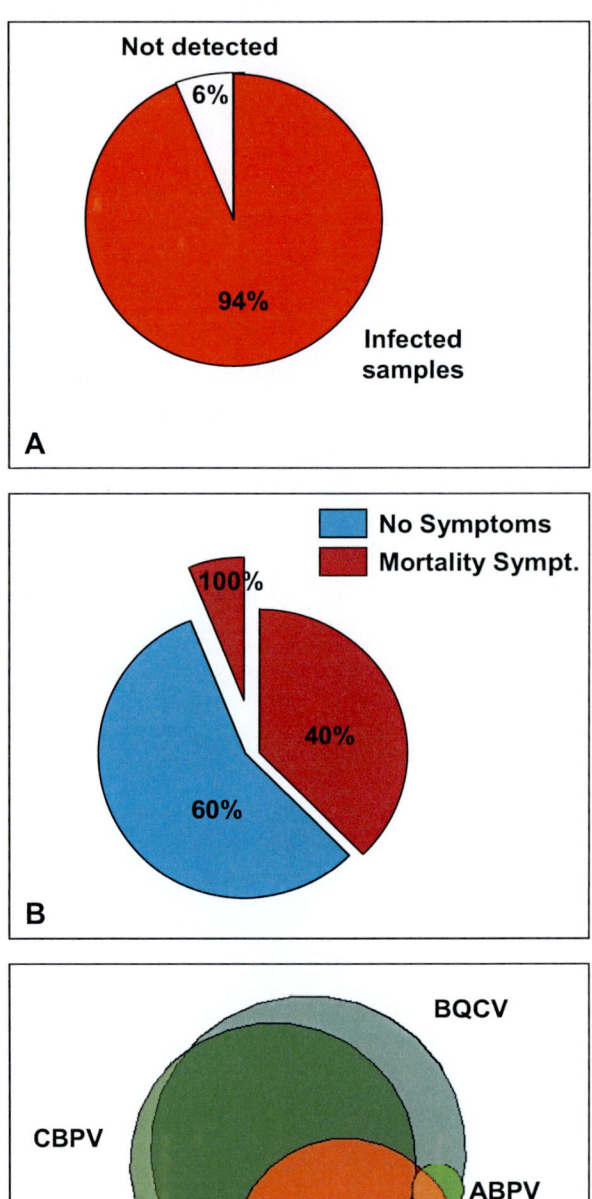

Figure 2. Analysis of the presence of CBPV, ABPV and BQCV in honeybees from Uruguay in a survey carried out during the winter of 2006.
A. Proportion of viral infection in the samples.
B. Distribution of observed symptoms.
C. Relations between the presence of different viruses and mortality symptoms.

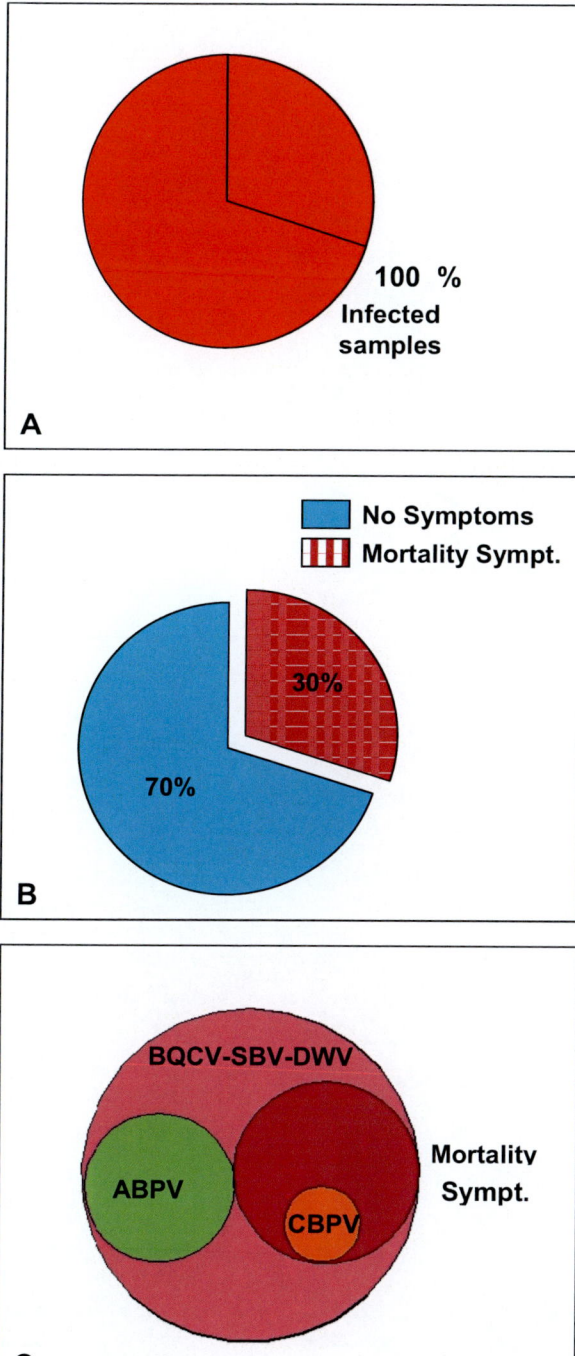

Figure 3. Analysis of the presence of CBPV, ABPV, BQCV, SBV and DWV in honeybees from Uruguay in a survey carried out during the summer of 2006.
A. Proportion of viral infection in the samples.
B. Distribution of observed symptoms.
C. Relations between the presence of different viruses and mortality symptoms.

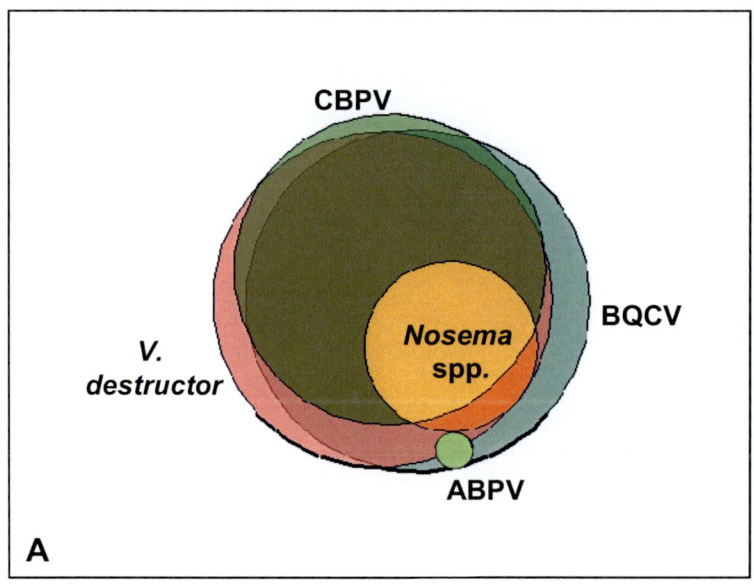

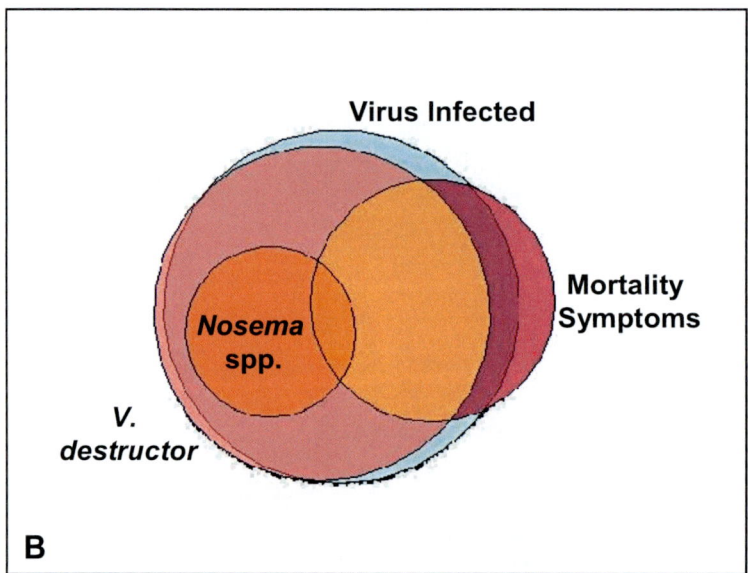

Figure 4. Analysis of the presence of *V. destructor* and *Nosema* spp. in honeybees from Uruguay in a survey carried out during the winter of 2006.
A. Relations between samples with different viruses, *V. destructor* and *Nosema* spp.*.
B. Relations between samples infected by viruses, *V. destructor* and *Nosema* spp. and mortality symptoms.
* The size of the "ABPV" circle was reduced from the original for better representation of the relations.

Nosema Spp. and *V. destructor* in Winter

V. destructor was present in 100 % of the analyzed samples and *Nosema* spp. in 28%. As all samples were infected by *V. destructor*, every sample was also infected or co-infected with viruses (Figure 4A). Subsequently, all samples infected with *Nosema* spp. were also infected with viruses, being most of them co-infected with BQCV and CBPV.

Most of the samples obtained from hives that showed mortality symptoms were co-infected by *V. destructor* and viruses. However, only one of these samples was also infected with *Nosema* spp. (Figure 4B).

Nosema Spp. and *V. destructor* in Summer

In the summer samples *V. destructor* infection was 98%, a similar value to that observed in the case of winter samples. *Nosema* spp. infection was higher than in the winter, reaching 65% and all the samples infected with *Nosema* spp. were also infected with *V. destructor*.

Accordingly to the high level of infection with *Nosema* spp. and *V. destructor*, several samples were additionally co-infected with ABPV, BQCV, SBV and DWV (Figure 5A). Finally, although all the samples with mortality symptoms were infected by viruses, this was not the case for samples infected by *Nosema* spp. or *V. destructor* because a group of the samples associated to honeybee mortality was not infected by these pathogens (Figure 5B).

Discussion

Viruses of the honey bee have been known for a long time, but during the last years have received special attention due to the honeybee mortality episodes that are taking place around the world. In order to evaluate the relation between the presence of viruses and bee mortality in Uruguay, as a first approach of the situation in South America, we evaluated the presence of different viruses during a two year survey. Surprisingly, five of the six analyzed viruses were present in Uruguay (ABPV, CBPV, BQCV, DWV and SBV). Most of the analyzed honeybee samples (from hives with or without mortality symptoms) were infected, indicating a high prevalence and wide distribution of viruses in Uruguay. ABPV, BQCV and DWV were also detected in Argentinean and Brazilian samples, indicating that these viruses are widely spread in South America [Antúnez, unpublished data, Teixeira et al., 2008]. KV was not detected in any of these countries, although it is present in North America, Europe, Asia and Oceania. However, its prevalence around the world is lower than that observed in the case of other viruses [Allen and Ball, 1996; Ellis and Munn, 2005; Siede et al., 2005].

According to the results shown in this chapter, the presence of symptoms associated to mortality of honeybees can be related to the presence of one or more RNA viruses in bees. This statement became more evident when more viruses were included in the analysis. When the six viruses were included in the analysis, in all the samples showing mortality symptoms at least one virus was detected. Although mortality symptoms could not be associated to one virus in particular, BQCV, SBV and DWV could be involved. However, the presence of viruses does not explain the development of symptoms by itself, since ABPV, CBPV, BQCV, SBV and DWV can persist in a latent state in beehives apparently healthy [Allen and Ball, 1996].

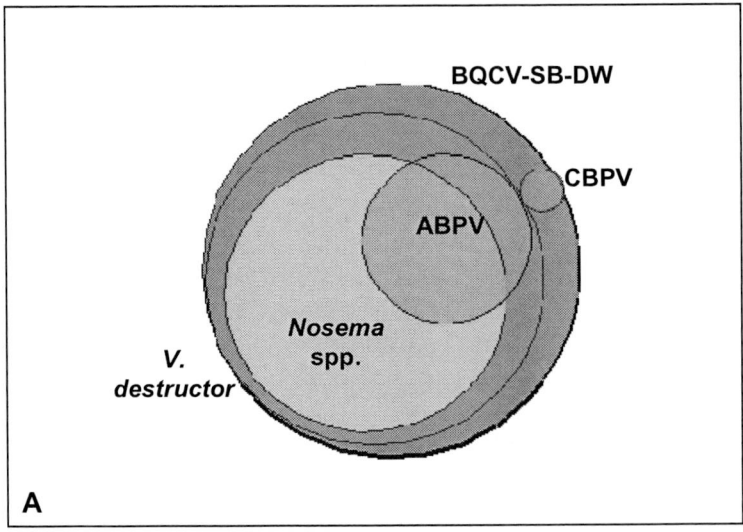

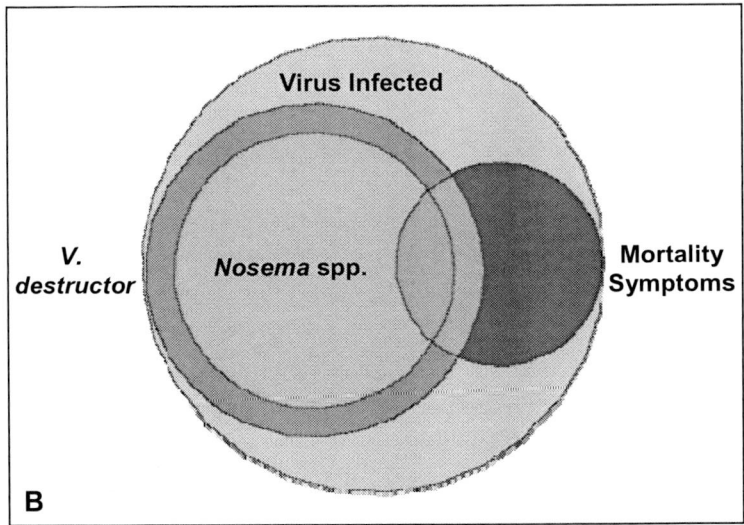

Figure 5. Analysis of the presence of *V. destructor* and *Nosema* spp. in honeybees from Uruguay in a survey carried out during the summer of 2006.
A. Relations between samples with different viruses, *V. destructor* and *Nosema* spp.*.
B. Relations between samples infected by viruses, *V. destructor* and *Nosema* spp. and mortality symptoms**.
*The size of the "CBPV" circle was reduced from the original for better representation of the relations.
**The size of the "Virus infected" circle was enlarged from the original for better representation of the relations.

The seasonal distribution of CBPV was clearly observed during both years, being its prevalence higher in winter than in summer. These findings oppose to previous reports which concluded that the presence of CBPV did not show a seasonal pattern [Tentcheva et al., 2004b]. In the case of ABPV we could not detected a seasonal behaviour, because in both years the infection rate per season was different, while Tentcheva et al. [2004b] found a higher prevalence in summer. The infection rate of BQCV was high during all year round, coinciding with previous reports [Tentcheva et al., 2004b].

The presence of viruses is not the only sanitary problem of Uruguayan honeybees. For example, *V. destructor* is a serious disease that beekeeping has to face every year. *V. destructor* has been present in Uruguay for decades and currently it is widely spread in the country. According to the results presented in this chapter, 98% of the analyzed samples were infected by this pathogen.

Another important honeybee pathogen that has gained special attention during the last years is *Nosema ceranae* [Higes et al., 2006]. It was usually considered that this pathogen exclusively colonized *Apis ceranae* until it was detected also in *A. mellifera*. Afterwards, different reports proposed that this pathogen is associated with episodes of honeybee mortality [Higes et al., 2006; Higes et al., 2007; Higes et al., 2008]. In our work, we did not differentiate *N. ceranae* from *N. apis*, but according to recent studies that are being conducted in our country (Invernizzi, personal communication), only *N. ceranae* is present in Uruguay. However, no relation between *V. destructor* or *Nosema* spp. and mortality symptoms was detected.

This study can be considered the first approach to understanding the relation between honeybee viruses and mortality in the region and could be taken into account for further analyses.

An important point to consider for the optimization of molecular detection of viruses is the choice of diagnostic technique. In the present chapter, we described the use of traditional RT-PCR, which is one of the most accurate methods for the diagnosis of RNA viruses, providing a quick, specific and sensitive procedure widely used to detect viruses in bees [Bakonyi et al., 2002; Benjeddou et al., 2001; Davison et al., 2003; Genersch, 2005; Grabensteiner et al., 2007; Grabensteiner et al., 2001; Parrella et al., 2006; Topley et al., 2005]. However, this technique does not allow the quantification of viruses, which could be important to elucidate the influence of the different viruses on honeybee mortality. One attractive alternative for this purpose is the use of Real Time RT-PCR, which is now being used by several researchers for the detection and quantification of different honeybee viruses [Chen et al., 2005; Chantawannakul et al.; 2006, Kukielka et al., 2008].

RNA stability is another important issue that has to be taken into account for molecular diagnosis of honeybee viruses, because the quality of RNA is one of the most critical components for the success of traditional and Real Time RT-PCR. Since honeybees must be collected in the field and delivered, RNA can be degraded, especially when colonies are situated far from the diagnostic laboratory. The lack of good-quality RNA can lead to false negative results. Chen et al. [2007] evaluated different sample-conservation methods, and they found that the storing of bees at –80°C or –20°C improved RNA quantity and preserved its integrity. They also reported that storing samples in liquid nitrogen or on dry ice could be the best method for the transport of samples.

Finally, it would be interesting to include the analysis of other important viruses in this kind of study, such as the Israeli acute paralysis virus (IAPV). This virus was first detected in honeybees from Israel in 2002 [Maori et al., 2007]. A metagenomic approach carried out in the United States in order to elucidate the factors implied in the mortality of honeybees proposed that IAPV was an important factor involved in this process [Cox-Foster et al.,

2007]. Also, a group of scientists detected a honeybee virus of 38 nm using the Integrated Virus Detection System. They proposed that it could be related to mortality of bees and are currently working on its identification [VDSC, 2008].

Conclusion

This chapter reports the first analysis of the presence of different RNA viruses and their relationship with the incidence of *V. destructor* and *Nosema* spp. in Uruguay and South America.

According to the results obtained in this study, it can be proposed that although the specific causes of honeybee mortality cannot be established, the presence of RNA viruses is closely related to mortality-associated symptoms in bees.

References

Allen, M. and Ball, B. (1996). The incidence and world distribution of honey bee viruses. *Bee World*, 77, 141-162.

Antúnez, K., D'Alessandro, B., Corbella, E. and Zunino, P. (2005). Detection of Chronic bee paralysis virus and Acute bee paralysis virus in Uruguayan honeybees. *Journal of Invertebrate Pathology*, 90, 69-72.

Antúnez, K., D'Alessandro, B., Corbella, E., Ramallo, G. and Zunino, P. (2006). Honeybee viruses in Uruguay. *Journal of Invertebrate Pathology*, 93, 67-70.

Bailey, L. (1967). The incidence of virus diseases in the honey bee. *Annals of Applied Biology*, 60: 43-48.

Bailey, L. (1975). Recent research on honey bee viruses. *Bee World*, 56, 55-64.

Bailey, L., and Woods, R.D. (1974). Three previously undescribed viruses from the honey bee. *Journal General Virology*, 25, 175-186.

Bailey, L., Gibbs, A. and Wood, R.D. (1963). Two viruses from adult honey bees (Apis mellifera Linnaeus) *Virology*, 21, 390-395.

Bakonyi, T., Farkas, R., Szendroi, A., Dobos-Kovacs, M. and Rusvai, M. (2002). Detection of acute bee paralysis virus by RT-PCR in honey bee and Varroa destructor field samples: Rapid screening of representative Hungarian apiaries. *Apidologie*, 33, 63-74.

Ball, B.V. and Bailey, L. (1997). Viruses. In R.A. Morse and K. Flottum (Eds), Honey, Bee Pest, Predators and Diseases (pp. 11-31). Medina, OH: A.I. Root Co.

Benjeddou, M., Leat, N., Allsopp, M. and Davison, S. (2001). Detection of Acute Bee Paralysis Virus and Black Queen Cell Virus from Honeybees by Reverse Transcriptase PCR. *Applied and Environmental Microbiology*, 67, 2384-2387.

Chantawannakul, P., Ward, L., Boonham, N. and Brown, M. (2006). A scientific note on the detection of honeybee viruses using real-time PCR (TaqMan) in Varroa mites collected from a Thai honeybee (Apis mellifera) apiary. *Journal of Invertebrate Pathology*, 91, 69-73.

Chen, Y., Evans, J., Hamilton, M. and Feldlaufer, M. (2007). The influence of RNA integrity on the detection of honey bee viruses: Molecular assessment of different sample storage methods. *Journal of Apicultural Research, 46,* 81-87.

Chen, Y.P., Higgins, J.A. and Feldlaufer, M.F. (2005). Quantitative real-time reverse transcription-PCR analysis of deformed wing virus infection in the honeybee (Apis mellifera L.). *Applied and Environmental Microbiology, 71,* 436-441.

Chen, Y.P. and Siede, R. (2007). Honey Bee Viruses. *Advances in Virus Research, 70,* 33-80.

Cox-Foster, D.L., Conlan, S., Holmes, E.C., Palacios, G., Evans, J.D., Moran, N.A., Quan, P.-L., Briese, T., Hornig, M., Geiser, D.M., Martinson, V., VanEngelsdorp, D., Kalkstein, A.L., Drysdale, A., Hui, J., Zhai, J., Cui, L., Hutchison, S.K., Simons, J.F., Egholm, M., Pettis, J.S. and Lipkin, W.I., (2007). A metagenomic survey of microbes in honey bee colony collapse disorder. *Science, 318,* 283-287.

Davison, S., Leat, N. and Benjeddou, M. (2003). Development of molecular tools for honeybee virus research: The South African contribution. *African Journal of Biotechnology, 2,* 867-878.

De Miranda, J.R., Drebot, M., Tylor, S., Shen, M., Cameron, C.E., Stolz, D.B. and Camazine, S.M. (2004). Complete nucleotide sequence of Kashmir bee virus and comparision with acute bee paralysis virus. *Journal of General Virology, 85,* 2263-2270.

Delaplane, S.K., and Mayer, F.D. (2000). *Crop pollinization in bees.* Cambridge, UK: University Press.

Ellis, J.D. and Munn, P.A.(2005). The worldwide health status of honey bees. *Bee World, 86,* 88-101.

Genersch, E., (2005). Development of a rapid and sensitive RT-PCR method for the detection of deformed wing virus, a pathogen of the honeybee (Apis mellifera). *Veterinary Journal, 169,* 121-123.

Ghosh, R.C., Ball, B.V., Willcocks, M.M. and Carter, M.J. (1999). The nucleotide sequence of sacbrood virus of the honey bee: An insect picorna-like virus. *Journal of General Virology, 80,* 1541-1549.

Govan, V.A., Leat, N., Allsopp, M. and Davison, S. (2000) Analysis of the complete genome sequence of acute bee paralysis virus shows that it belongs to the novel group of insect infecting RNA viruses. *Virology, 277,* 457-463.

Grabensteiner, E., Bakonyi, T., Ritter, W., Pechhacker, H. and Nowotny, N. (2007) Development of a multiplex RT-PCR for the simultaneous detection of three viruses of the honeybee (Apis mellifera L.): Acute bee paralysis virus, Black queen cell virus and Sacbrood virus. *Journal of Invertebrate Pathology, 94,* 222-225.

Grabensteiner, E., Ritter, W., Carter, M.J., Davison, S., Pechhacker, H., Kolodziejek, J., Boecking, O., Derakhshifar, I., Moosbeckhofer, R., Licek, E. and Nowotny, N. (2001). Sacbrood virus of the honeybee (Apis mellifera): Rapid identification and phylogenetic analysis using reverse transcription-PCR. *Clinical and Diagnostic Laboratory Immunology, 8,* 93-104.

Higes, M., García-Palencia, P., Martín-Hernández, R., Meana, A. (2007). Experimental infection of Apis mellifera honeybees with Nosema ceranae (Microsporidia). *Journal of Invertebrate Pathology, 94,* 211-217.

Higes, M., Martín, R. and Meana, A. (2006). Nosema ceranae, a new microsporidian parasite in honeybees in Europe. *Journal of Invertebrate Pathology, 92,* 93-95.

Higes, M., Martín-Hernández, R., Botías, C., Bailón, E.G., González-Porto, A.V., Barrios, L., Del Nozal, M.J., Bernal, J.L., Jiménez, J.J., Palencia, P.G. and Meana, A. (2008). How natural infection by Nosema ceranae causes honeybee colony collapse. *Environmental Microbiology.* In press.

Kestler, H.A., Müller, A., Kraus, J.M., Buchholz, M., Gress, T.M., Liu, H., Kane, D.W., Zeeberg, B.R. and Weinstein, J.N. (2008). VennMaster: area-proportional Euler diagrams for functional GO analysis of microarrays. *BMC Bioinformatics, 9,* 67.

Kukielka, D., Esperón, F., Higes, M. and Sánchez-Vizcaíno, J.M. (2008). A sensitive one-step real-time RT-PCR method for detection of deformed wing virus and black queen cell virus in honeybee Apis mellifera. *Journal of Virology Methods, 147,* 275-281.

Lanzi, G., De Miranda, J.R., Boniotti, M.B., Cameron, C.E., Lavazza, A., Capucci, L., Camazine, S.M. and Rossi, C. (2006). Molecular and biological characterization of deformed wing virus of honeybees (Apis mellifera L.). *Journal of Virology, 80,* 4998-5009.

Leat, N., Ball, B., Govan, V. and Davison, S. (2000). Analysis of the complete genome sequence of black queen cell virus, a picorna like virus of honey bees. *Journal of General Virology, 81,* 2111-2119.

Maori, E., Lavi, S., Mozes-Koch, R., Gantman, Y., Peretz, Y., Edelbaum, O., Tanne, E. and Sela, I., (2007). Isolation and characterization of Israeli acute paralysis virus, a dicistrovirus affecting honeybees in Israel: Evidence for diversity due to intra- and inter-species recombination. *Journal of General Virology, 88,* 3428-3438.

Morse, R.A. and Flottum, K. (1997). *Honey Bee Pests Predators and Diseases.* Medina, OH: A.I. Root Co.

Morse, R.A. and Calderone, N.W. (2000). The value of honey bees as pollinators of U.S. crops in 2000. *Bee Culture, 128,* 2-15

Parrella, G., Caprio, E. and Mazzone, P. (2006). Development of improved molecular methods for the detection of deformed wing virus (DWV) in honeybees (Apis mellifera L.) and mites (Varroa destructor Oud.). *Communications in agricultural and applied biological sciences, 71,* 625-629.

Ribiere, M., Tribolout, C., Mathieu, L., Aurieres, C., Faucon, J. P. and Pepin, M. (2002). Molecular diagnosis of chronic bee paralysis virus infection. *Apidologie, 33,* 339-351.

Schmid-Hempel P. (1998). Parasites in social insects. Princeton University Press, Princeton, NJ.

Sambrook, J., Fritsch, E. F. and Maniatis, T. (1989). Molecular Cloning. A laboratory manual, Cold Spring Harbor Laboratory Press, NY.

Siede, R., Derakhshifar, I., Otten, C., Berenyi, T., Koglberger, H., and Buchker, R. (2005). Prevalence of Kahmir bee virus in Europe. *Journal of Apicultural Research, 44,* 131-132.

Stoltz, D., Shen, X. R., Boggis, C., and Sisson, G. (1995). Molecular diagnosis of Kashmir bee virus infection. *Journal of Apicultural Research, 34,* 153-160.

Suchail, S., Debrauwer, L. and Belzunces, L.P. (2004). Metabolism of imidacloprid in Apis mellifera. *Pest Management Science, 60,* 291-296.

Teixeira, E.W., Chen, Y., Message, D., Pettis, J. and Evans, J.D. (2008). Virus infections in Brazilian honey bees. *Journal Invertebrate Pathology.* In press.

Tentcheva, D., Gauthier, L., Jouve, S., Canabady-Rochelle, L., Dainat, B., Cousserans, F., Colin, M. E., Ball, B. V. and Bergoin, M. (2004a). Polymerase chain reaction detection of deformed wing virus (DWV) in Apis mellifera L., and Varroa destructor. *Apidologie. 35*, 431- 439.

Tentcheva, D., Gauthier, L., Zappulla, N., Dainat, B., Cousserans, F., Colin, M.E. and Bergoin, M. (2004b). Prevalence and seasonal variations of six bee viruses in Apis mellifera L. and Varroa destructor mite populations in France. *Applied and Environmental Microbiology, 70*, 7185-7191.

Topley, E., Davison, S., Leat, N. and Benjeddou, M. (2005). Detection of three honeybee viruses simultaneously by a single Multiplex Reverse Transcriptase PCR. *African Journal of Biotechnology, 4*, 763-767.

Virus Detection System Corporation (2008). Detection of unknown viruses extracted from bee samples. http://www.vdsc.us/documents/analyzed_viruses.pdf.

In: Insect Viruses: Detection, Characterization and Roles
Editors: Ch. J. Connell and D. P. Ralston
ISBN: 978-1-60692-965-0
© 2009 Nova Science Publishers, Inc.

Chapter 5

Encephalitic Arboviruses: Emerging and Re-Emerging Problem

Agostino Pugliese[*], *Tiziana Beltramo*[1] *and Donato Torre*[2]

[1]Department of Medical and Surgical Sciences, Section of Clinical Microbiology of Turin University, "Amedeo di Savoia" Hospital, Corso Svizzera, Turin, Italy

[2]Section of Infectious Diseases, General Hospital, Cittiglio, Varese, Italy

Abstract

At present Arboviruses include 534 viruses of them 134 are of human interest, and a large part of the last ones are endowed with neurotropism. In particular, this ecological classification concerns different families and viral genera, but all the viruses included in the same classification, are characterized by transmission through arthropod bites. These viruses can be responsible for human infections ranging from asymptomatic or mild ones to fatal illness, like encephalitis and hemorrhagic fevers. The most important arboviruses causing human pathology are included in three viral families, Togaviridae (genus Alphavirus), Flaviviridae, and Bunyaviridae.

Different distribution of encephalitic arboviruses exists among various areas of the world, however generally these viruses are present, in large or small degree, in all the temperate or warm climate regions, and sometimes also in more cold areas (but in this case only in the hot season). The diffusion of these infections parallels the cycle of their vectors: especially mosquitoes, ticks and phlebotomes. Habitat modification induced by men, as deforestation, can change the ecology of vectors and the epidemiology of encephalitides transmitted by arboviruses, but also of other vector–borne diseases. This phenomenon is due both to climate and to demographic changes, and in particular occurs in South America, West and Central Africa and in South-East Asia. Moreover also the

---

[*] Corresponding author: Agostino Pugliese agostino.pugliese@unito.itPersonal address: Prof. Agostino Pugliese Via Polonghera, 9, 10138 – Torino, Italy.

rapidity of modern transports has contributed to encephalitic arboviruses new localization areas that sometimes are more adapted to cause major epidemic phenomena. The modification in reservoir host environment and the viral agents adaptation to new hosts, represent the main cause of pattern modification of these infections in the world. West Nile virus is a classic example of this phenomenon. In fact, recently, this virus was introduced in the Western Hemisphere. Consequently arboviruses can no more be considered specific of particular regions of the world, but represent a more wide potential danger for men. Since the acquired immunity of population against some arboviruses and the lability in the external environment of a large part of them may risk their extinction, the evolution has adopted a dual host tropism strategy for viral survival: reservoir hosts and final, or occasional hosts. Men just sometimes represent the last ones that may have clinical manifestations.

In order to prevent these serious pathologies, because there are no specific therapies for arboviral encephalitides, fight against vectors and vaccines employ can constitute the only defences for these diseases. For example effective vaccines exist against Tick-borne encephalitis virus (TBEV) and Japanese encephalitis virus (JEV).

In conclusion in our chapter we analyse the general questions concerning the arboviral encephalitides in the world (biological characteristics of the causative agents, epidemiology, pathogenicity, diagnosis and prophylaxis) and in the last part of this paper we analyse the Italian situation. In fact, beginning from 2000 we have studied the epidemiology of arboviruses in Piedmont and in particular in the Turin Province. Finally we report also some data concerning arboviral biology.

Agostino Pugliese is Professor of Clinical Microbiology at the Medical Faculty of Turin University. Dr. Donato Torre is Specialist of Infectious Diseases, and of Paediatrics.

Introduction

At present Arboviruses (Arthropod-borne viruses) include 534 viruses of them 134 are of human interest, and in large part are endowed with neurotropism (Gubler, 2001; 2002). In particular, this ecological classification concerns different families and viral genera, but all the viruses included in the same classification, are characterized by transmission through arthropod bites (Bres, 1980). However a few authors have included in this classification also viruses not transmitted by arthropods, like as Rabdoviridae (Rehle, 1998). Human Arboviruses can be responsible of infections ranging from asymptomatic or mild ones to fatal illness, like as encephalitis, or encephalomyelitis, and hemorrhagic fevers (Rehle, 1989). The most important arboviruses causing human pathology are included in three viral families, Togaviridae (genus Alphavirus), Flaviviridae, and Bunyaviridae (Nalca et al., 2003). Nevertheless, some viruses with the characteristics of human arboviruses are classified also in other viral groups; for instance some encephalitic viruses of old and new world are included in the genus Coltivirus (Family Reoviridae – Attoui et al., 2002). The causative agent of Colorado tick fever is an example of the last ones (Romero and Newland, 2003).

In particular, Dengue haemorrhagic fever, yellow fever and Japanese encephalitis are generally considered the most important arboviral diseases of the men (Rehle, 1989).

As regard the arboviral encephalitides, about twenty are the most diffused ones, and its distribution can sometimes interest not only rural, but also urban areas (see for pathognomonic example West Nile virus infections, but also La Crosse, and Saint Louis encephalitides) (Zeller, 1999; Rust *et al.*, 1999; Reisen, 2003).

Different distribution of encephalitic arboviruses exists among various areas of the world, however generally these viruses are present, in large or small degree, in all the temperate or warm climate regions, and sometimes also in more cold areas (but in this case only in the hot season). In fact, for example, Tick borne encephalitis was described also in Central Siberia (Kislenko *et al.*, 1997), and in cooler areas of Eastern Europe (Solomon and Mallewa, 2001). Besides Jamestown Canyon encephalitis virus (California serogroup of Buniaviridae family), and Tahyna, Inkoo and Batai viruses (Bunyavirus genus of Bunyaviridae family) are present in Central Russia (Walters *et al.*, 1999; Korobeinikova *et al.*, 2003; Vanlandingham *et al.*, 2002) and in Alaska (Walters *et al.*, 1999).

The diffusion of arboviral encephalitis infections parallels the cycle of their vectors: especially mosquitoes, ticks and phlebotomes, and often is affected by climatic and seasonal conditions (Mellor and Leake, 2000). Consequently climate change can influence the emergence and re-emergence of arthropods transmitted infections, and particularly the mosquito-borne ones (Patz *et al.*, 1996). On this subject, an increase of global temperature of 2°C is expected by the year 2100 (Patz *et al.*, 1996). Moreover habitat modification induced by men, as deforestation, can change the ecology of vectors and the epidemiology of encephalitides transmitted by arboviruses, but also of other vector–borne diseases (Walsh *et al.*, 1993). This phenomenon is due both to climate changing, and to demographic modifications, and in particular occurs in South America, West and Central Africa and in South-East Asia (Walsh *et al.*, 1993). For instance the necessity of introducing rice cultivation in deforested areas, especially to nourish Asiatic population, facilitates mosquito's diffusion (Gajanana *et al.*, 1997). Moreover the modification of farming organization or the abandonment of scarcely producing agricultural areas can favour tick-borne infections (Mekonnen *et al.*, 2002). Besides the rapidity of modern transports has contributed to encephalitic arboviruses new localization areas that sometimes are more adapted to cause major epidemic phenomena (Gubler, 2001). Finally the changes in reservoirs behaviour and arboviral agents' adaptation to new hosts, are relevant causes of pattern modification of these infections in the world. West Nile virus is a classic example of this phenomenon (Gubler, 2001; Mc Lean *et al.*, 2002). In fact, recently, this virus was introduced in the Western Hemisphere, conforming to local reservoirs of infection. Consequently arboviruses cannot be more considered specific of particular regions of the world, but they represent a wide potential danger for men (Nalca *et al.*, 2003). Since the acquired immunity of population against some arboviruses and the lability in the external environment of a large part of them may risk their extinction, the evolution has adopted a dual host tropism strategy for viral survival: reservoir and terminal hosts (often occasional) (Peterhans *et al.*, 1999). Men just can constitute the last ones that may have clinical manifestations, but also asymptomatic or paucisymptomatic infections. In particular men can act as occasional and terminal hosts (Powell and Kappus, 1978), with the exception of yellow fever, dengue, Sandfly fevers and few other arboviral infections. Often in arboviral endemic areas a high seroprevalence rate is present in the population and asymptomatic forms prevail on symptomatic ones (Romano-Lieber and Iversson, 2000).

Different animals can contribute to Arboviruses diffusion. In particular they can act as hosts of maintenance (reservoirs), occasional and terminal (generally no have significance in the viral transmission), of connection (among reservoirs and terminal hosts) or virulence enhancer ones, and able to infection amplify (Hassan, et al., 2003). Besides the reservoirs can be wild or domestic vertebrates (mammals, birds, sometimes reptiles and amphibians). However, also some invertebrate vectors may sometimes act as viral reservoirs (Rodhain, 1998). In the last case it is also possible the viruses transovarian transmission or the viral persistence, during the cold season, in hibernant species (Rodhain, 1998).

Generally small vertebrates present a low, but persistent viremia, instead of big animals that have high and transient viremic levels; consequently the first category of hosts have a more efficient function in the maintenance of the arboviral infections in nature, than the second ones (Rodhain, 1998).

Finally, it is also important to recall that the transmission of arboviruses without involvement of arthropod vectors is also possible for animals and sometimes for men too (Kuno, 2001). In the case of ticks transmitted encephalitides the contagion is also possible through infected milk ingestion, especially of ovine and goat (Dumpis et al., 1999).

In order to prevent arboviral encephalitides, vaccines, and the fight against the vectors (e.g. use of N,N-diethyl-m-toluamide, DEET = 20% in the Autan; Permethrin and Citronellal – Cockcroft et al., 1998) can represent the only defences for these diseases, because of the lack of specific therapies. Effective vaccines are usually employed for at risk population, included travellers, against Tick-borne encephalitis virus (TBEV) and Japanese encephalitis virus (JEV) (Nalca et al., 2003). The last virus at time seems to be numerically the most important cause of epidemic encephalitis in the world (Solomon and Mallewa, 2001). Instead the dengue infections, that only sometimes may produce neurological complications, i.e. aseptic meningitis, or more serious manifestations, constitute the most diffused arboviral diseases (Solomon. and Mallewa, 2001). In this last case vaccine studies are in progress, but with caution for the risk of increasing the immuno-mediated haemorrhagic manifestations (Chang et al., 2004). Moreover also for equine encephalitides, Rift Valley fever, St. Louis encephalitis, California and Murray Valley encephalitides, vaccines are available or in progress, but also in the former case, sometimes these preparations are furnished with difficult (Nalca et al., 2003).

Despite the absence of specific therapy for viral encephalitides, with the exception of the herpetic ones and of very few others, an early diagnosis is useful for a non-specific intensive or support therapies, when they are requested. Often, focal neurological manifestations arisen suddenly, suggest the onset of viral encephalitis, more rarely a generalised picture appears until the beginning. (Chaudhuri and Kennedy, 2002). However a biphasic behaviour is generally a common characteristic of these infections, although prodromic manifestations are often aspecific and sometimes not recognized (Kaiser, 2002).

American Epidemiology of Arboviral Encephalitides

A different speech is suitable for North and Latin America. In fact, in the last prevail the equine encephalitides (main vectors mosquitoes *Aedes* and *Culex* species), so named because produce epidemic diseases among the horses that may be followed by human epidemics (Lhuillier *et al.*, 1981). These encephalitides include: Eastern equine encephalitis (EEE), Western equine encephalitis (WEE) and Venezuelan equine encephalitis (VEE) complex (family Togaviridae, genus Alphavirus). These infections are prevalently present in the eastern regions the first, in the western the second and especially in Venezuela, Peru, Columbia, Brazil and Mexico the last one (Fernandez *et al.*, 2003). However the causative agents of equine encephalitides that have as reservoirs wild and domestic birds, rodents, some snakes and amphibians, are generally sporadic in the USA, but sometimes are periodically present in the Southern States (Przelomski *et al.*, 1988; Calisher, 1994). Moreover during 1975-77 in the Ribeira Valley (Brazil), occurred also a further encephalitic epidemic caused by a new flavivirus named Rocio (Iversson *et al.*, 1989).

In North America the most important encephalitic arboviruses are: St. Louis encephalitis agent (SLE – Flaviviridae, Flavivirus) a serious encephalitis also of historic interest, having as main vectors some species of *Culex* mosquitoes and reservoirs wild birds; Powassan virus (POWV – Flavivirus), transmitted by ticks and similar to Euro-Asiatic TBE virus (main reservoirs, small rodents); and La Crosse virus (LACV – Bunyaviridae, Bunyavirus). LACV has as vectors: mosquitoes of *Ochlerotatus* and *Aedes* species, and as main reservoirs small rodents (Calisher, 1994). Instead Jamestown Canyon virus (JVC – Bunyavirus of California group - transmitted by mosquitoes and having as reservoirs hares and deer), responsible of sporadic cases of encephalitis in the North of USA and in Canada (Mayo *et al.*, 2001) and Colorado tick fever (CTF) virus (Reoviridae, Coltivirus), are less aggressive agents (Calisher, 1994; Mancao *et al.*, 1996; Charrel *et al.*, 1999; Alciati *et al.*, 2001; Nalca *et al.*, 2003). In particular, the last virus, of them the squirrels are the main reservoirs, generally causes only a benign fever that is present in North Western USA and in Canada. Fortunately, in this case, only rarely CNS is reached, prevalently with aseptic meningitis involvement (Klasco, 2002).

Recently the epidemiologic picture of American arboviruses has been complicated by the occurrence of encephalitides induced by West Nile virus (WNV) in the USA (Emig and Apple, 2004). This virus, prevalently transmitted by various species of *Culex*, affects also big mammals, and in the USA the crows are their main reservoirs (Smith and Fonseca, 2004). In particular in 2002 the United States were affected by a dramatic epidemic of West Nile virus disease that produced 4156 cases of documented neurologic infection, and 284 patients died (overall mortality rate of 6.8%, that significantly increased in subjects $\geq$ 65 years). Monoparesis prevailed in $\leq$ 50 years patients, whereas paraparesis or quadriparesis in those aged $\geq$ 65 years (Emig and Apple, 2004).

Coming back to the subject of equine encephalitides, the Venezuelan one, caused serious epidemic episodes during the 1969-72 in Central America and in these occasions, the vaccination attempt were sometimes responsible of neurological disease, because of incomplete inactivation of the infective agent (Brown, 1993). These episodes produced fear for a possible employ of this virus or of other encephalitic arboviruses as a potential biological weapon (Schlesinger, 2001; Han and Zunt, 2003). Also Easter Equine Encephalitis virus has produced widespread epidemic in Latin America, or sporadic in North one, with a mortality much higher than VEE. In particular, Deresiewicz *et al.* (1997) detected in the USA (1988-1994) a mortality of 36% and in 35% of survivors, described moderate or severe disability. However, mortality rate progressively declined in the course of the years, especially if compared with the seventy ones (Letson *et al.*, 1993). These epidemics of equine encephalitis can be correlated with excess rainfall (Letson *et al.*, 1993), and genetic and antigenic diversity were detected among EEE viruses from North, Central and South America (Brault *et al.*, 1999). Moreover it has been demonstrated that WEEV, a more benign virus than EEEV, derived from a recombination between the last one and Sindbis virus (Hahn *et al.*, 1988).

St.Louis virus has not only a great historic interest (Tsai *et al.*, 1987) (epidemics of half twentieth century), but also a more recent one, because of a worrying re-emergence episode. In fact, 63 cases of St. Louis encephalitis were described in Northeast Louisiana during 2001 (Jones *et al.*, 2002). The most cases affected subjects among 45 years and older and were clinically characterised by fever, meningitis syndrome, tremors and altered mental status. The mortality rate was about of 4.5% (Jones *et al.*, 2002). Previously in the summer of 1996, an epidemic was described in Corpus Christi, Texas (76 cases confirmed, with two deaths = 2.6% mortality rate). An epidemic was described the same year also in Dallas; this episode followed an epidemic, broken out in Huston, during 1964 (Williams *et al.*, 1975).

Powassan is a prevalent American virus, among tick-borne group encephalitic flaviviruses that include also the Malaysian Langat virus, and not American Louping ill and TBE viruses. As previously recalled, Powassan virus is like as TBEV, however produces encephalitis on a smaller scale than the last virus in Euro-Asiatic areas (Gritsun *et al.*, 2003 a,b).

La Crosse encephalitis virus is at present endemic in Western North Carolina (Utz *et al.*, 2003). Lifelong neurologic sequels affect about 20% of the encephalitic patients. The name of this illness derived from a historic epidemic that affected the children that played in the neighbours of La Crosse city, when deforestation produced puddles that favoured the mosquito's development. In particular in Mid-Atlantic and Midwestern USA States an incidence of 20-30/100,000 inhabitants was described in the last 1900 years. This rate exceeded the incidence of bacterial meningitis (Mc Junkin *et al.*, 1998).

Other less important American encephalitic arboviruses are: Everglades = EVE, Florida (Alphavirus) transmitted by mosquitoes; Ilheus = ILH, Central and Southern America (Flavivirus), transmitted by mosquitoes; Tensaw = TEN, North America (Bunyavirus) transmitted by mosquitoes; Snowshoe hare = SSH, USA, Canada, bu also Siberia (Bunyavirus), transmitted by mosquitoes (Mitchell *et al.*, 1996; Figuereido, 2000).

Table 1. Main encephalitic arboviruses of American Group

Viruses	Vectors	Main reservoirs	Main diffusion lands	Lethality
Eastern equine encephalitis - EEE (Alfavirus, Togaviridae)	Mosquitoes	Birds Bats	USA Atlantic coasts, Gulf of Mexico, Caribe, Eastern sides of Latin Americas (EEE reached also Philippinas isles, Thailand and Centro-Eastern Europe)	About 35%, but also sometimes even > 50%
Western equine encephalitis - WEE (Alfavirus)	Mosquitoes	Birds, Snakes Wild rodents	West States of USA, West sides of Centre and part of South America.	About 5%
Venezuelan equine encephalitis- VEE (Alfavirus)	Mosquitoes	Mammals, also of large size, Birds, Snakes, Frogs	South States of USA, especially Florida and Texas, Mexico, Central Americas and in some northern areas of South America	<5%
Saint Louis encephalitis - SLE (Flavivirus - Flaviviridae)	Mosquitoes	Birds, Mammals, Hibernating mosquitoes	Almost all USA States, and Caribe	Even to ≥ 20%, but generally lower (5-10%)
Californian encephalitis Group (Bunyavirus, Bunyaviridae); the most important of them are: California encephalitis - CE La Crosse-LAC and Jamestown Canyon encephalitis - JC	Mosquitoes	Mammals, Mosquitoes (by transovarial modality)	California, but also other States of USA, especially in the case of La Crosse virus that is endowed with more wide diffusion. California serogroup was recently diffused in Asia	<5%
Powassan-POW (Flavivirus)	Ticks	Birds, Mammals	New York State, North USA, California, Canada	<5%
West Nile – WN (Flavivirus)	Mosquitoes	Birds	Recently reached USA, Canada and Central America	2-12% (variable, according to different lands)

African Epidemiology of Arboviral Encephalitides

About twenty arboviruses pathogenic for humans are present in the Central Africa. The most important are West Nile, Yellow Fever and Rift Valley Fever (Mathiot et al., 1988). In particular, Yellow fever, mosquitoes-transmitted (*Aedes aegypti* in urban form), is an infective hepatitis with characteristics of generalized, and haemorrhagic infection that can involve also Central Nervous System (CNS) (Schlesinger *et al.*, 1996); moreover, also Rift Valley Fever is a systemic infection with haemorrhagic manifestations and possible CNS involvement (Siam *et al.*, 1980; Madani *et al.*, 2003). Mosquito's bite represents the main transmission modality of the virus that however can infect the hosts also by inhalation or alimentary way (Balkhy and Memish, 2003). Bats constitute the probable reservoir hosts (Oelofsen and Van der Ryst, 1999). Virus diffusion is associated to a hard increase of its mosquito vectors, related to periods of high rainfall, especially in South Senegal (Bicout and Sabatier, 2004). Besides West Nile is a febrile infection that originated from West Nile District of Uganda, where the etiological agent was primarily isolated in 1937. The causative agent increased in virulence during the following years and also acquired a marked neurovirulence (Deubel *et al.*, 2001).

Instead in Maghreb, a part West Nile Virus (WNV), prevail three tick-borne arboviruses: Soldado (Nairovirus, Bunyaviridae), Essaouria and Kala Iris (Orbivirus, Reoviridae – Kemerovo Serogroup), that however generally cause only limited febrile illness, and are present also Sandfly fever viruses and Toscana virus, that are diffused in all Mediterranean areas (Chastel *et al.*, 1995). Moreover in southern Algeria West Nile virus has been isolated since 1968, and in Tunisia other than WNV was isolated also a new Phlebovirus ticks-transmitted, responsible of febrile ill, and named Tunis virus (Chastel *et al.*, 1995). However in Nile River Valley of Egypt more important arboviral risk is present. In fact, a seroepedemiologic study of Corvin *et al.* (1993) demonstrated in the adult population a seroprevalence of 6% for Sandfly fevers, of 15% for Rift Valley fever and of 20% for West Nile viruses (Corvin *et al.*, 1993).

West Nile virus is present not only in Mediterranean areas, but also in Central and South Africa, where the virus is maintained in a wide biological cycle involving feral and domestic birds, humans or other mammals (terminal hosts) and ornithophilic mosquitoes (Jupp, 2001). In particular two strains lineages were found: lineage 1 from Central and North Africa, that involved also Europe, Israel, and North America; lineage 2 from Central and Southern Africa and Madagascar. Other than encephalitis also cases of fatal hepatitis has been described in humans affected by WNV (Burt *et al.*, 2002).

In the last twenty years serotype 3 (subtype III) of dengue virus has caused wide epidemics of hemorrhagic fever in East Africa, other than in Sri Lanka and Latin America. This viral variant originated from the Indian subcontinent and reached the Africa in the 1980s (Messer *et al.*, 2003). It is interesting to underline that occasionally neurologic manifestations can occur also during dengue fever (Patey *et al.*, 1993). In this case the most frequent clinical manifestations are reduced consciousness and convulsions, moreover neurological sequels can be also possible (Solomon *et al.*, 2000).

In particular, it is important to recall that dengue causes 60 millions of cases, and 30,000 deaths/year in the world. The infection is transmitted by the bite of *Aedes* mosquitoes that seem to be also the main reservoirs of the disease (Tolou *et al.*, 1997). However bats too seem to be involved in dengue virus transmission, aging as reservoirs (Scott, 2001).

Finally Thogoto (THO) virus (Orthomyxoviridae), has been described particularly in Egypt, Kenya, Nigeria, Uganda and Central African Republic, as not frequent cause of encephalitis (Ogen-Odoi *et al.*, 1999). Eight strains of this virus were isolated in Africa, Asia and Europe (Kuno *et al.*, 2001).

Table 2. Main encephalitic arboviruses of Africa

Viruses	Vectors	Main reservoirs	Main diffusion lands	Lethality
West Nile - WN(Flavivirus, Flaviviridae)	Mosquitoes	Birds	Northern Africa, but also various areas of Central and Southern Africa, included Madagascar	2-12% (variable, according to different lands)
Toscana Virus – TOSV (Phlebovirus, Bunyaviridae)	Phlebotomes	Birds ? Mammals ?	Mediterranean basin	< 5%
Rift Valley - RV (Flavivirus)	Mosquitoes	Rodents (bats) Ruminants ?	Egypt, Mauritania, Central and East Africa, South Africa (occasionally encephalitis)	About 15%, but even 36% in case of haemorrhagic syndr. with hepatitis and nephritis
Thogoto - THO (Orthomyxoviridae)	Ticks	Rodents (mangooses)	Egypt, Nigeria Kenia, Central African Rep., Congo, Uganda	?

Euro-Asiatic Epidemiology of Arboviral Encephalitides

One of the most important arboviral encephalitis of Euro-Asiatic group is tick-borne one (TBE), Russian and Centro-European type. In fact, this form of encephalitis in Europe and Asia, involves approximately 15,000 annual cases in men, especially in Russia (Gritsun *et al.*, 2003 a,b). Moreover in the only Sweden 40-130 cases of meningo-encephalitis, TBEV-induced, are found each year, all belonging to the Western TBEV subtype. The same is also widespread in Austria, or in other states of Central Europe, and is less virulent than Eastern subtype (Russian subtype) (Haglund *et al.*, 2003). In reality, in the case of these encephalitides it is more exact to speak of TBE complex that consists in fourteen viral strains antigenically correlated. Eight of them are pathogen for men. In particular, the strains Neudorfl (Western subtype) and Sofjin (Eastern subtype) are considered all belonging to two subtypes of the same virus, transmitted respectively by ticks *Ixodes ricinus* and *Ixodes persulcatus*, that constitute also important viral reservoirs, together with some species of rodents, and sometimes of big mammals too, as sheep, deer and wild boars. Western subtype is responsible of tick borne encephalitis of Central Europe, and Eastern subtype of Russian spring-summer encephalitis, that reached also confining states. The last form shows a more severe pattern that the former one (see review of Alciati *et al.*, 2001). Instead, tick-transmitted related viruses, as Louping ill, and Powassan, rarely are the cause of human encephalitis on epidemic scale in Euro-Asiatic area (Gritsun *et al.*, 2003 a). In particular, Powassan virus is of particular interest for the North America, but it is present also in Russia in sporadic form (Leonova *et al.*, 1991), and Louping ill is an epizootic encephalitis that interests particularly the North England and Ireland, but very rarely affects the men (Davidson *et al.*, 1991).

Several mosquito-transmitted arboviruses (sometimes encephalitic) are present in Europe. The main vectors belong to *Aedes*, *Anopheles* and *Culex* species (Lundstrom, 1999). One of the most important is the neurotropic West Nile virus (Flaviviridae) that was isolated especially in Southern and Central Europe: France, Portugal, Romania (especially during 1996), Czechoslovakia, Russia (Lundstrom, 1999). This virus is maintained in African endemic foci from which can reach the South Europe and Middle East (McLean *et al.*, 2002). During the years 1996-2001 the spreading of illness produced by WNV, in animal and humans, widely increased. In particular encephalitis was described in horses in Italy and France (1998-2000) and in domestic birds in Israel (1997-99). In 1999 WNV reached the USA, and humans, horses and also crows were affected by mortal illness (Mc Lean *et al.*, 2002).

Among the Euro-Asiatic arboviruses several are Bunyaviridae, belonging to the California group. Tahyna and Inkoo viruses (perhaps the most important), but also other neurotropic arboviruses are of interest; for instance Bhanja, Erve, Kemerovo group and Eyach, Thogoto and Batai viruses (Dobler,1996; Lundstrom, 1999). In particular, Tahyna virus, responsible of respiratory symptoms and sometimes also of central nervous system impairment, has been described in Central and Southern Europe, and is transmitted prevalently by *Aedes* mosquitoes (Demikhov and Chaitsev, 1995). Besides Inkoo virus, *Aedes*-transmitted too, may cause encephalitis in Northern Europe, including the Russia (Lundstrom, 1999).

Dengue is also present in Euro-Asiatic areas and is transmitted particularly by *Aedes albopictus*, a mosquito species originating from Asia, but successively diffused in all tropical and temperate regions of the world, occupying the ecological habitat of *Aedes aegypti* (Rodhain, 1996).

An interesting seroepidemiologic study about arboviruses was made in 1996 on travellers returning in Germany from Mediterranean areas or from the tropics. In this study, seropositivity for Toscana Virus, Dengue, and other minor arboviruses was sometimes detected (Schwarz *et al.*, 1996).

Table 3. Main encephalitic arboviruses of Euro-Asiatic Group

Viruses	Vectors	Main reservoirs	Main diffusion lands	Lethality
Russian Spring – Summer encephalitis - TBE (Flavivirus, Flaviviridae)	Ticks (goat milk as vehicle)	Mammals, Birds, Ticks	Russia and confining States – Eurasian forests	>30% in encephalitic forms and especially in encephalomyelitic ones
Central European encephalitis- TBE (Flavivirus)	Ticks (goat milk as vehicle)	Mammals, Birds, Ticks	Especially Central Europe and Danubio Basin. In particular, was described in Czech Republic, Poland, Austria, Germany, Unghery, French, Switzerland, Jugoslavy, Italy	Generally <5%, but in the rare case of encephalomyelitis, also even >20%
Louping ill - LI (Flavivirus)	Ticks (Manipulation of infected animals can also transmit the infection to men)	Mammals, Birds, Ticks	England, Ireland.	<10%
Kemerovo - KEM (Orbivirus, Reoviridae)	Ticks	Rodents	Central Europe.	<10%
Tahyña (Bunyavirus)	Mosquitoes	Mammals	Czech Republic, Italy, Jugoslavy, Southern French	<10%
WestNile, WNV (Flavivirus)	Mosquitoes	Birds	Northern Africa, Middle East (especially Egypt and Israel), Eastern Europe, Russian, USA.	2 - 12% (variable, according to different lands)
Toscana Virus – TOSV (Phlebovirus, Bunyaviridae)	Phlebotomes	Birds ? Mammals (rodents)?	Mediterranean basin (especially Italy, Portugal, Spain and Cypro).	<5 %

Main Encephalitic Arboviruses of Far East Group

In particular, Toscana virus was described the first time in Italy, in Toscana, in early years '80 by Prof. Paola Verani of Virology Laboratory (Istituto Superiore di Sanità, Roma) (Verani *et al.*, 1984). This virus is a new member of Phlebotomus fever serogroup (family Bunyaviridae, genus Phlebovirus), related to Naples and Sicilian Sandfly fever viruses (Verani *et al.*, 1984), but widely diffused in the Mediterranean basin (especially Italy, Portugal, Spain, Cyprus, and Middle East) (Valassina *et al.*, 2003). Toscana virus can be responsible of meningitis, meningoencephalitis and encephalitis that increase their incidence in the summer months, and their severity in old patients (Valassina *et al.*, 2003). In the case of the only meningitis, that generally is benign, headache is present in all patients, moreover a moderate fever of mean duration of 48 h is present in about ¾ of the patients and nuchal rigidity in half of them (Navarro *et al.*, 2004)

Japanese encephalitis (JE) and Murray Valley (MV) viruses are the most important neurotropic agents of this group. In particular JEV is diffused especially in Southeast Asia (Kalita *et al.*, 2003), but also in Northern Australia and Papua New Guinea (Spicer, 2003; Van Den Hurk *et al.*, 2003), in Japan (Ma *et al.*, 2003), and Russia (Loginova and Karpova, 2001). Movement disorders and personality alterations are possible in subjects affected by this encephalitis. The mortality rate is major in adults than in children patients and also the sequels are more evident in the first ones, because of the lesser neuronal plasticity (Kalita *et al.*, 2003). In India a serious epidemic of JEV broke out during 2000, with a fatality rate of 42.11%. This epidemic reached the peak in the monsoon months of July and August and had as main vectors the mosquitoes *Culex vishnui* and *Mansonia annulifera* (Kaur *et al.*, 2002). Climate variability correlated with JEV transmission was described also in China (Bi *et al.*, 2003). Birds, pigs (feral too), rodents and rarely marsupials can constitute the reservoir hosts of JEV (Van Den Hurk *et al.*, 2003); however also horses may be infected by the virus and so contribute to encephalitis diffusion (Widjaja *et al.*, 1995).

Murray Valley encephalitis virus (MVEV) is particularly present in the basin of Murray River, but was described also in Southwest, and in Northwest regions of Australia (Broom *et al.*, 2003; Burrow *et al.*, 1998). The vectors are various species of *Culex* and reservoirs particularly wild or domestic birds, but perhaps also feral pigs (Gard *et al.*, 1976). An increased incidence of the infection is in part related with rainfalls and floodings (Broom *et al.*, 2003). Burrow *et al.* (1998) described a mortality of 31% among the patients admitted for MVE to Royal Darwin Hospital from 1987-1996, and in a quarter of cases continued to be a residual neurologic disability.

Moreover an other important encephalitic arbovirus is present in Australia: Kunjin virus (Flavivirus) that often coexists with Ross River virus, not encephalitic alphavirus, responsible of fevers with polyarthralgias (Johansen *et al.*, 2003), and with Murray Valley encephalitis virus (Brown *et al.*, 2002). These viruses can affect also Australian horses (Studdert *et al.*, 2003). In particular the two over-mentioned encephalitic Australian arboviruses are endemic in the tropical part of Northern and Western Australia, but till to 2000, from 1974 were absent in Central Australia. During 2000 an encephalitic epidemic followed unexpected high rainfall in this dry area. This re-emerging epidemic focus represents a potential risk for reintroduction of two viruses into South-eastern Australia areas at time not affected (Brown *et al.*, 2002).

Table 4. Main encephalitic arboviruses of Far East Group

Viruses	Vectors	Main reservoirs	Main diffusion lands	Lethality
Japanese encephalitis - JE (Flavivirus, Flaviviridae)	Mosquitoes	Mammals, Birds	Russian Far East, Japan, Chorea, China, Taiwan, Philippinas isles, Indonesia, Northern Australia, Papua New Guinea	About 40%, but sometimes even >50%
Negishi - NEG (Flavivirus)	Ticks	Mammals, Birds	Japan, Eastern Russia	<10%
Kyasanur Forest Complex - KFD (Flavivirus)	Ticks	Mammals, Birds	India	<10%
Murray Valley encephalitis- MVE (Flavivirus)	Mosquitoes	Mammals, Birds	Australia, in particular the basin of Murray River.	Even > 30%

Kyasanur Forest disease virus (KFDV) is a distinct member of tick-borne encephalitis group, originating from the India, that like as Omsk virus, produces especially haemorrhagic manifestations (Venugopal *et al.*, 1994). However, during 1959 in Karnataka State (India) occurred a shift from hemorrhagic to neurologic complications of the infection (Pavri, 1989). KFDV is transmitted by *Ixodid* ticks, besides rodents and sheep are its reservoirs (Saxena, 1997).

Finally Negishi virus was firstly described in Eastern Russia in 1950 (Honda, 1951) and was identified as a new member of Russian spring-summer encephalitis agents that reached Japan (Okuno *et al.*, 1961). However more recently Negishi and Louping ill (a zoonotic agent of England and Ireland) viruses were found to be most closely related to Sofjin strain of TBE subgroup (Shamanin *et al.*, 1991).

East equine encephalitis may be included in part in this group, because was signalled in Far East and in particular in Philippinas isles, in Thailand and in Indonesia.

Clinical Manifestations

Fortunately, often encephalitic arboviruses infections are asymptomatic, especially among the population living in the endemic areas (Bres, 1980; Powell and Kappus, 1979). For example TBE virus, of Centro-European subtype is asymptomatic in 70% of cases, instead in remaining 30%, generally, only influenza-like symptoms are present as headache, fever (often also of high degree), arthromyalgias, pharyngitis and gastro-enteric manifestations, that compared after an incubation period of 2-28 days (mean 14 days).

Generally these clinical signs endure only 2-4 days, but in 10-30% of the cases biphasic behaviour is present with subsequent CNS involvement, that follows an apparent remission. Neurologic manifestation can be: aseptic meningitis (that can also spontaneously regress), encephalitis, meningoencephalitis, and sometimes encephalomyelitis (Kaiser, 2002). The severity of infection is proportional to the age of patients and the mortality rate is 1-3%, but not rarely neurologic sequels occur in survivors (Kaiser, 2002; Jereb et al., 2002, Grygorczuk et al., 2002).

Spring - summer TBE Russian subtype has a mortality rate higher (from 20 to 60% of encephalitic cases), and a more rapid onset with more severe symptoms and sequels, that can affect also over 50% of survivors. In particular, Leonova et al., 1987 described in Russia's far-eastern Primorski Territory, a mean average of 50% of focal forms and a mortality of 6.4% for meningeal forms and of 33% for encephalitic ones.

In reality, also other arboviral encephalitides have a similar picture with a more or less severe evolution, and the biphasic behaviour is a common clinical characteristic.

Headache, emesis, photophobia, rigidity of nape and rachis are common signs of meninges compromising, but if the evolution leads into meningoencephalitis or encephalomyelitis, appear deliquium, paralysis, convulsions, ataxia. In these cases, if the patients survive, paralysis, epileptic manifestations, parestesias and personality disturbance can remain as sequels that are more frequent in old patients.

Also in the case of St Louis encephalitis a higher severity paralleled the increase of patients age, and a fatality rate of 35-38% have been reported for patients aged ≥ 60 years, during an important epidemic of 1975 that involved 1791 cases (Powell and Kappus, 1978).

Eastern equine encephalitis, the most serious among the equine encephalitides, has a mortality rate of about 35%, but sometimes also higher, and about 35% of survivors present moderate or severe disability (Deresiewicz et al., 1997). The most frequent neuroradiographic abnormalities, especially detected with magnetic resonance imaging (MRI), included focal lesions in the basal ganglia, and in the thalamus, followed by brain stem lesions. The detection of early involvement of basal ganglia and thalami is relevant for differential diagnosis with herpes simplex encephalitis (Deresiewicz et al., 1997). During an epidemic of Venezuelan equine encephalitis broke out in 1995 in Venezuela, 313 patients affected by encephalitis were hospitalised (about 3% of 11,072 patients medically treated for the epidemic). Intracranial hypertension was the main clinical sign (about 56% of the patients), but the mortality rate was relatively moderate = 1.7% (Molina et al., 1999).

Murray Valley encephalitis, as described by Burrow et al. (1998), in the Australian Northern Territory, produced often in adults spinal cord and brainstem involvement, with a mortality of 31%, and neurological disability in 25% of patients evaluated.

Japanese encephalitis have an overall case fatality rate of about 42% (Kaur et al., 2002), and the sequels are more frequent in adults than in children (Potula et al., 2003). Potula et al., 2003, that described a lower mortality rate (35.8%), reported as main symptoms: fever = 100% (212/212 patients studied); altered sensorium = 87.7%; convulsion = 85.8%, headache = 50%, and vomiting = 47.6%.

Deep studies were performed on tick-borne encephalitides etiopathogenesis. Experimental infections in animals demonstrated that tick-borne encephalitis group viruses induce neuronal cells death, both directly by viral replication, and indirectly by the attacks of immune system (Kenyon *et al.*, 1992). This occurs also in the case of other encephalitic arboviruses, of them the replication possibility in NCS, depends on the ability to cross the haemato-encephalic barrier. These viruses can affect all nervous structures, including basal nuclei, cerebral cortex, cerebellum, brain stem and spinal cord. Moreover also radiculoneuritis is possible, generally of immuno-mediated etiology.

Neuronal cells degeneration is associated to neuronophagy and microglia proliferation. Besides also cell death for apoptosis may be relevant, not only in the case of neuronal cells, but also in glial cells (Garen *et al.*, 1999). Moreover the mechanism by which encephalitic arboviruses can reach CNS is now again discussed, even if two possible mechanisms are the most probable: growth across endothelial cells of brain capillaries, and passive diffusion in consequence of the high viremic levels. However also olfactory pathway is sometimes possible, as occurs in the case of viral aerosol production (see for instance laboratory infections) (Monath *et al.*, 1983).

Diagnosis

Laboratory investigations, that include direct and indirect tests are fundamental for arboviral encephalitides etiological diagnosis.

A classic direct method is the isolation from the liquor (not from the blood, because of the viremia shortness) or from bioptic and autoptic specimens, despite the risk connected for the laboratorians (Vernet, 2004). Other direct methods are immunofluorescence, and immunoenzimatic tests, or more recent PCR techniques.

Indirect methods include complement fixation, neutralization, haemoagglutination-inhibition test, indirect immunofluorescence, ELISA tests and also immunochromatographic techniques. The last two diagnostic systems can be employed not only for the serum evaluations, but also for CSF investigations and permit also the easy detection of specific IgM in the sera (Vernet, 2004). However a limit of serologic tests is the possible existence of cross-reactions among different arboviruses. Moreover biochemical and cytological evaluation of liquor, EEG, cranial computed tomography and magnetic resonance imaging, can also be employed for an initial diagnostic approach or for the detection of viral induced damages (Morse *et al.*, 1992; Kaiser, 2002).

Ixoes ricinus　　　　　　　　　　　　　　　　　　　Aede albopictus

Figure 1. Two examples of arthropods vectors of human infections.

Personal Contribution

Seroprevalence Study of Tick Borne Encephalitis, Dengue and Toscana Virus in Turin Province

As previously reported Tick borne encephalitis (TBE) is one of the most severe infections transmitted by ticks (Alciati *et al.*, 2001). The endemic area spreads from the Rhine to the Urals, from Scandinavia to Italy and Greece (Jaussaud *et al.*, 2001). In Southern Europe, the western European type is more common than the Russian type.

In the Italy, where the main vector of TBE is *Ixodes ricinus* (Verani *et al.*, 1979), the diffusion of TBE in the last ten years is significantly increased because of the progressive abandonment of human installations in the country sides, and especially in the pre-alpine areas. This favoured ticks diffusion in confining areas with those endemic for TBE (Bassetti et al., 1992; Verani *et al.*, 1995). Moreover, *Aedes albopictus*, potential vector of many arboviruses spreads in Italy since 1990 (Romi, 2001), together with many other vectors of arboviral infections.

Seroepidemiological studies about TBE were performed in Trentino and Tuscany in subjects exposed to the risk of tick's bite (Jaussaud *et al.*, 2001; Verani *et al.*, 1995; Cristofolini *et al.*, Hudson *et al.*, 2001). According to Bassetti *et al.* (1992), hunters are the subjects mainly exposed to TBE in Trentino (2.1% of seroprevalence), followed by hunt wardens (1.3% of seroprevalence).

These data and the lack of information about TBE epidemiology in Piedmont (Italy) led us to analyse with a specific ELISA test the sera of 3 different categories of subjects at risk for tick's bite. The categories studied consisted in: i) hunters from country-side surrounding Turin, ii) wild boar breeders from Susa Valley and iii) adults with a particular interest in outdoor activities in the country side. This third group of subjects was considered at moderate or low risk for TBE infection. Our study group is the first that analysed the sera from subjects exposed to TBE risk in Piedmont.

In parallel we analysed also Dengue and Toscana virus seroprevalence in the same subjects because of the possible presence of dengue sporadic cases, generally related to travellers, in Italy (Mele *et al., 2001*), and because the diffusion of Toscana virus in some areas of Central Italy (Dionisio *et al.*, 2003).

Table 5 reports the seroprevalence for TBEV, TOSV and Dengue in our subjects, detected with immunoenzymatic methods (Toscana Virus = DIESSE, Siena, Italy; Dengue = DRG, Marburg, Germany; TBE = Pantec S.r.l., Torino, Italy).

For cases significantly positive for TBE antibodies were found among the at risk population (hunters and wild boar breeders). Moreover, three of the significantly positive subjects referred in their anamnesis manifestations that could be correlated to TBE previous infection, such as tiredness, fever under 39°C, headache and arthromialgias, however with absence of neurologic signs. Only one borderline case was detected in the group of subjects with a moderate or low risk. Besides no cases of seropositivity for Lyme disease, a borreliosis, also transmitted by tick's bits, were found in our study. This datum is not reported in table because our chapter treats only arboviral encephalitides. Instead considering a population of 40 subjects at moderate or low risk of arthropod bits, collected during 1993-95, no positive cases were detected for all the three viruses studied.

Table 6 reports the means of antibody titres for TBE in the groups studied during year 2001 and in comparison with a moderate/low risk group of '93-95.

A significant difference was found comparing, by ANOVA test, the groups at high risk, when evaluated separately or together (hunters + breeders), with the group of subjects at moderate-low risk for tick's bite of 2001 ($p < 0.001$ in the case of hunters, $p < 0.001$ in that of breeders and $p < 0.001$, considering hunters and breeders together). Moreover using Student's t – test a significant difference for $p = 0.01$ was detected between hunters and wild boar breeders, demonstrating that the last ones are less exposed to tick's bite than the formers, and for $p < 0.001$ comparing the mean value of moderate low risk subjects of 1993-95 with the mean value of hunters + breeders together.

Finally a study was performed in order to detect how much two important factors of aspecific immunity, as body temperature and Interferon (IFN) production, could be related to behaviour of Sindbis virus, a typical arbovirus, that often is employed as experimental model of research for arboviral encephalitides (Griffin, 1998). This study was performed in order to explain the role of reservoirs and hosts on arboviruses persistence in nature. Table 7 reports the effect of different incubation temperature on Sindbis replication, in chick embryo cells, and Table 8, the IFN production induced in human and animals cells by Sindbis virus.

Table 5. Percentage of seropositivity and borderline response against three arboviruses in Turin and Province

Viruses	Subjects	Positive	Borderline
TBEV	At risk	5.7	2.8
	Moderate– low risk	0	2.5
DENGUE	At risk	2.0	2.0
	Moderate–low risk	0	0
TOSV	At risk	2.0	4.0
	Moderate–low risk	2.5	0

Table 6. Means anti TBE virus antibody titres (U/ml) in different categories of subjects studied in Piedmont

Subjects	ELISA test titre
Hunters	80.4±45.6
Wild boar breeders	53.5±32.8
Hunters + Breeders	66.9±35.1
Moderate/low risk subjects (year 2001)	21.5±10.5
Moderate/low risk Subjects (years '93-95)	22.4±5.4

Significant difference for $p < 0.001$ by ANOVA test.

The over reported data suggest that the body temperature could be determinant as regard the differentiation between reservoir or possible symptomatic host, but the interferon production seems not be implicated in this phenomenon.

Finally our data demonstrate the presence of TBE virus in Piedmont.

Table 7. Effect of temperature incubation on Sindbis virus replication *in vitro*

Incubation temperature	Haemagglutination titre (U/ml)
25°C	45.4±10.2
37°C	341.3±120.6
39°C	85.3±15.1

Significant difference for $p < 0.001$ by ANOVA test.

Table 8. Interferon production induced *in vitro* by Sindbis virus

Cell line	IFN titer (UI/ml)
WISH	37.0±7.5
L-929	32.0±11.3
ECF	34.6±9.9

No significant differences into the groups by ANOVA test.

The Table reports the interferon (IFN) values produced by Sindbis virus on cell lines derived from three different species: human, WISH (epithelial-like amniotic cells); murine, L-929 (fibroblasts); chicken, ECF (embryonal chicken fibroblasts).

The spread of TBE virus in Piedmont, like as occurs in other areas of Central and Northern Italy (Cristofolini *et al.*, 1993; Verani *et al.*, 1995; Hudson *et al.*, 2001), seems to involve only subjects highly exposed to tick's bite risk. This situation is different from other European regions, in particular Middle West European lands, where the infection can be contracted also by individuals that occasionally go into the areas where the risk for tick 'bite is high.

In particular, TBE virus epidemiologic condition in the Susa Valley, derived from the wild boar breeders studied, seems to be very similar as Trentino epidemiology, which is today considered the most important area for what concerns TBE virus spread in Italy. Moreover, this Region is right next to Austria, where the risk of TBE virus infection is known to be very high (Pisecky and Freund, 2003). However, the studies regarding TBE virus infection in Trentino were performed both on subjects at moderate risk for tick's bite (such as occasional hunters), and on subjects at high risk for tick's bite (such as wild boar breeders). This was difficult, in a wide way, in our study because of the limited number of hunters, or wild boar breeders who agreed to undergo our seroepidemiologic investigations.

In conclusion, we believe that wider studies to investigate the occurrence of TBE virus infection in Piedmont area could be useful in order to protect the subjects that are exposed to tick's bite risk. These studies could also be able to explain the unknown aetiology of some cases of encephalitis in this geographical area and might constitute a useful suggestion for similar studies in other areas where arboviral encephalitides could be also present.

References

Alciati S., Belligni E., Del Colle S., Pugliese A. Human infections tick-transmitted. *Panminerva Med.* 2001; 43: 295-304.

Attoui H, Mohd Jaafar F, Biagini P, Cantaloube JF, de Micco P, Murphy FA, de Lamballerie X. Genus Coltivirus (Family Reoviridae) genomic and morphologic characterization of Old World and New World viruses. *Arch Virol.* 2002; 147: 533-61.

Balkhy HH, Memish ZA. Rift Valley fever: an uninvited zoonosis in the Arabian peninsula. *Int J Antimicrob Agents* 2003; 21: 153-57.

Bassetti D., Ciufolini MG., Nicoletti L., Verani P. Indagine sieroepidemiologica sulla diffusione del virus Tick Borne Encephalitis in Trentino. *Microbiol Med.* 1992; 7: 143-45.

Bi P, Tong S, Donald K, Patron KA, Ni J. Climate variability and transmission of Japanese encephalitis in eastern China. *Vector Borne Zoonotic Dis.* 2003; 3: 111-15.

Bicout DJ, Sabatier P. Mapping Rift Valley Fever vectors and prevalence using raifall variations. *Vector Borne Zoonotic Dis.* 2004; 4: 33-42.

Brault AC, Powers AM, Chavez CL, Lopez RN, Cachon MF, Gutierrez LF, Kang W, Tesh RB, Shope RE, Weaver SC. Genetic and antigenic diversity among eastern equine encephalitis viruses from North, Central, and south America. *Am J Trop Med Hyg.* 1999; 61: 579-86.

Bres P. Arboviral diseases: a field where research and public health call for international cooperation. *Med Trop.* 1980; 40: 485-91.

Broom AK, Lindsay MD, Wright AE, Smith DW, Mackenzie JS. Epizootic activity of Murray Valley encephalitis and Kunjin viruses in an aboriginal community in the southeast Kimberley region of western Australia: results of mosquito fauna and virus isolation studies. *Am J Trop Med Hyg.* 2003; 69: 277-83.

Brown A, Bolisetty S, Whelan P, Smith D, Wheaton G. reappearance of human cases due to Murray Valley encephalitis virus and Kunjin virus in central Australia after an absence of 26 years. *Commun Dis Intell.* 2002 ; 26 : 39-44.

Brown F. Review of accidents caused by incomplete inactivation of viruses. *Dev Biol Stand.* 1993; 81: 103-107.

Burrow JN, Whelan PI, Kilburn CJ, Fisher DA, Currie BJ, Smith DW. Australian encephalitis in the Northern Territory: clinical and epidemiological features, 1987-1996. *Aust N Z J Med.* 1998; 28: 590-96.

Burt FJ, Grobbelaar AA, Leman PA, Anthony FS, Gibson GV, Swanepoel R. Phylogenetic relationships of southern African West Nile virus isolates *Emerg Infect Dis.* 2002; 8: 820-26.

Calisher CH. Medically important arboviruses of the United States and Canada. *Clin Microbiol Rev.* 1994; 7: 89-116.

Chang GJ, Kuno G, Purdy DE, Davis BS. Recent advancement in flavivirus vaccine development. *Expert Rev Vaccines* 2004; 3: 199-220.

Charrel RN, Levy N, Tesh RB, Chandler LJ. Use of base excision sequence scanning for detection of genetic variations in St. Louis encephalitis virus. *J Clin Microbiol.* 1999; 37: 1935-40.

Chastel C, Bailly-Choumara H, Bach-Hamba D, De Lay G, Legrand MC, Le Goff F, Vermeil C. Tick-transmitted arbovirus in Magreb. *Bull Soc Pathol Exot*. 1995; 88: 81-85.

Chaudhuri A, Kennedy PG. Diagnosis and treatment of viral encephalitis. *Postgrad Med J*. 2002; 78: 575-83.

Cockcroft A, Cosgrove JB, Wood RJ. Comparative repellency of commercial formulations od deet, permethrin and citronellal against the moqsquito Aedes aegypti, using a collagen membrane technique compared with human arm tests. *Med Vet Entomol*. 1998; 12: 289-94.

Corwin A, Habib M, Watts D, Darwish M, Olson J, Botros B, Hibbs R, Kleinosky M, Lee HW, Shope R. *et al*. Community based prevalence profile of arboviral, rickettsial, and Hantaan-like viral antibody in the Nile River Delta of Egypt. *Am J Trop Med Hyg*. 1993; 48: 776-83.

Cristofolini A., Bassetti D., Schallenberg G. Zoonoses transmitted by ticks in forest workers (tick borne encephalitis and Lyme borreliosis): preliminary results. *Med Lav*. 1993; 84: 394-402.

Davidson MM, Williams H, Macleod JA. Louping ill in man: a forgotten disease. *J Infect*. 1991; 23: 241-49.

Demikhov VG, Chaitsev VG. Neurologic characteristics of diseases caused by Inkoo and Tahyna viruses. *Vopr Virusol*. 1995; 40: 21-25.

Deresiewicz RL, Thaler SJ, Hsu L, Zamani AA. Clinical and neuroradiographic manifestations of eastern equine encephalitis. *N Engl J Med*. 1997; 336: 1867-74.

Deubel V, Fiette L., Gounon P., Drouet MT, Khun H, Huerre M, Banet C, Malkinson M, Despres P. Variations in biological features of West Nile viruses. *Ann NY Acad Sci*. 2001; 951: 195-206.

Dionisio D, Esperti F, Vivarelli A, Valassina M. Epidemiological and laboratory aspects of sandly fever. *Curr Opin Infect Dis*. 2003; 16: 383-88.

Dobler G., Arboviruses causing neurological disorders in the central nervous system. *Arch Virol Suppl*. 1996; 11: 33-40.

Dumpis U, Crook D, Oksi J. Tick-borne encephalitis. *Clin Infect Dis*. 1999 ; 28 : 882-90.

Emig M, Apple DJ. Severe West Nile virus disease in healthy adults. Clin *Infect Dis*. 2004 ; 38 : 289-92.

Fernandez Z, Moncayo AC, Carrara AS, Forattini OP, Weaver SC. Vector competence of rural and urban strains of Aedes (Stegomyia), albopictus (Diptera: Culicidae) from Sao Paulo State, Brazil for IC, ID and IF subtypes of Venezuelan equine encephalitis virus. *J Med Entomol*. 2003; 40: 522-27.

Figuereido LT. The Brazilian flaviviruses. *Microbes Infect*. 2000; 2: 1643-49.

Gajanana A., Rajendran R, Samuel PP, Thenmozhi V, Tsai TF, Kimura-Kuroda J, Reuben R. Japanese encephalitis in south Arcot district, Tamil Nadu, India: a three-year longitudinal study of vector abundance and infection frequency. *J Med Entomol*. 1997; 34: 651-59.

Gard Gp, Giles R, Dwyer-Grey RJ, Woodroofe GM. Serological evidence of inter-epidemic infection of feral pigs in New South Wales with Murray Valley encephalitis virus. *Aust J Exp Biol Med Sci*. 1976; 54: 297-302

Garen PD, Tsai TF, Powers JM. Human eastern equine encephalitis: immunohistochemistry and ultrastructure. *Mod Pathol*. 1999; 12: 646-52.

Griffin DE. A review of alphavirus replication in neurons. *Neurosci Biobehav Rev*. 1998; 22: 721-23.

Gritsun TS, Lashkevich VA, Gould EA. Tick-borne encephalitis. *Antiviral Res*. 2003; 57: 129-46 (a).

Gritsun TS, Nuttall PA, Gould EA. Tick-borne flaviviruses. *Adv Virus Res*. 2003; 61: 317-71(b).

Grygorczuk S, Mierzynska D, Zdrodowska A, Zajkowska J, Pancewicz S, Kondrusik M, Swierzbinska R, Pryszmont J, Hermanowska-Szpakowicz T. The course of tick-born encephalitis (TBE) in patients hospitalized at the Department of Infectious Diseases in Biallystok in the year 2001. *Przegl Epidemiol*. 2002; 56: 595-604.

Gubler DJ. Human arbovirus infections worldwide. *Ann N Y Acad Sci*. 2001; 951 : 13-24.

Gubler DJ. The global emergence/resurgence of arboviral diseases as public health problems. *Arch Med Res*. 2002; 33: 330-42.

Haglund M, Vene S, Forsgren M, Gunter G, Johansson B, Niedrig M, Plysnin A, Lindquist L, Lundkvist A. Characterisation of human tick-borne encephalitis virus from Sweden. *J Med Virol*. 2003; 71: 610-21.

Hahn CS, Lustig S, Strauss EG, Strauss JH. Western equine encephalitis virus is a recombinant virus. *Proc Nat Acad Sci* USA. 1988; 85: 5997-6001.

Han MH, Zunt JR. Bioterrorism and the nervous system. *Curr Neurol Neurosci Rep*. 2003; 3: 476-82.

Hassan HK, Cupp EW, Hill GE, Katholi CR, Klingler K, Unnasch TR. Avian host preference by vectors of eastern equine encephalomyelitis. *Am J Trop Med Hyg*. 2003; 69: 641-47.

Honda Y. Comparison of the immunological characters of a new virus "Negishi strain" and Russian spring-summer encephalitis virus. *Kitasato Arch Exp Med*. 1951; 24: 29-34.

Hudson PJ., Rizzoli A., Rosa R., Chemini C., Jones LD., Gould EA. Tick borne encephalitis in northern Italy. *Med Vet Entomol*. 2001; 15: 304-13.

Iversson LB, Travassos da Rosa AP, Rosa MD. Recent occurrence of human infection by Rocio arbovirus in the Valley of Ribeira region. *Rev Inst Med Trop Sao Paulo*. 1989; 31: 28-31.

Jaussaud R., Magy N., Strady A., Dupond JL., Deville JF. Tick borne encephalitis. *Rev Med Interne* 2001; 22: 452-58.

Jereb M, Muzlovic I, Avsic-Zupanc T, Karner P. Severe tich-borne encephalitis in Slovenia: epidemiological, clinical and laboratory findings. *Wien Klin Wochenschr*. 2002; 114: 623-26.

Johansen CA, Nisbet DJ, Zborowski O, Van Den Hurk AF, Ritchie SA, Mackenzie JS. Flavivirus isolation from mosquitoes collected from western Cape York Peninsula, Australia, 1999-2000. *J Am Mosq Control Assoc*. 2003.;19: 3392-96.

Jones SC, Morris J, Hill G, Alderman M, Ratard RC. St. Louis encephalitis outbreak in Louisiana in 2001. *J La State Med Soc*. 2002; 154: 303-306.

Jupp PG. The ecology of West Nile virus in South Africa and the occurrence of outbreaks in humans. Ann NY Acad Sci. 2001; 951: 143-52.

Kaiser R. Tick-borne encephalitis (TBE) in Germany and clinical course of the disease. *Int J Med Microbiol*. 2002; 291 (Suppl. 33): 58-61.

Kalita J, Misra UK, Pandey S, Dhole TN. A comparison of clinical and radiological findings in adults and children with Japanese encephalitis. *Arch Neurol.* 2003; 60: 1760-64.

Kaur R, Agarwal CS, Das D. An investigation into the JE epidemic of 2000 in Upper Assam – a perspective study. *J Commun Dis.* 2002 ; 34 : 135-45.

Kenyon RH, Rippy MK, McKee KT Jr, Zack PM, Peters CJ. Infection of Macaca radiata with viruses of the tick-borne encephalitis. *Microb Pathog.* 1992; 13: 399-409.

Kislenko GS, Korotkov IS, Chunikhin SP. The results of the surgical examination of medium-size mammals in the natural foci arbovirus infections in central Siberia. *Med Parazitol.* 1997; 4: 28-32.

Klasco R. Colorado tick fever. *Med Clin North Am.* 2002; 86: 435-40.

Korobeinikova AS, Nafeev AA, Skvottsova TM. Specific markers for the detection of circulating Tahyna, Inkoo and Batai viruses (Bunyaviridae, Bunnyavirus) in humans, mosquitoes, ticks and cattle of the Ul'ianovsk region. *Vopr Virusol.* 2003; 48: 45-46.

Kuno G. Transmission of arboviruses without involvement of arthropod vectors. *Acta Virol.* 2001; 45: 139-50.

Kuno G, Chang GJ, Tsuchiya KR, Miller BR. Phylogeny of Thogoto virus. *Virus Genes* 2001; 23: 211-14.

Leonova GN, Rybachuk VN, Krugliak SP, Guliaeva SE, Baranov NI. Clinico-epidemiological characteristics of tick-borne encephalitis in the Maritime Territory. *Zh Mikrobiol Epidemiol Immunobiol.* 1987; 12: 40-45.

Leonova GN, Sorokina MN, Krughiak SP. The clinico-epidemiological characeristics of Powassan encephalitis in the southern Soviet Far East. *Zh Mikrobiol Epidemiol Immunobiol.* 1991; 3: 35-39.

Letson GW, Bailey RE, Pearson J, Tsai TF. Eastern equine encephalitis (EEE): a description of the 1989 outbreak, recent epidemiologic trends, and the association on rainfall with EEE occurrence. *Am J Trop Med Hyg.* 1993; 49: 677-85.

Lhuillier M., Pajot FX, Mouchet J, Robin Y. Arbovirus diseases in South America and Caribbean Islands. *Med Trop (Mars).* 1981 ; 41 : 73-84.

Loginova NV, Karpova EF. Genetic analysis of Japanese encephalitis virus strains and variants from Russian collection. *Vopr Virusol.* 2001; 46: 46-47.

Lundstrom JO. Mosquito-borne viruses in western Europe: a review. *J Vector Ecol.* 1999; 24: 1-39.

Ma SP, Yoshida Y, Makino Y, Tadano M, Ono T, Ogawa M. Short report : a major genotype of Japanese encephalitis virus currently circulating in Japan. *Am J Trop Med Hyg.* 2003; 69: 151-54.

Madani TA, Al-Mazrou YY, Al-Jeffri MH, Nishkhas AA, Al-Rabead AM, Turkistani AM, Al-Sayed MO, Abodahish AA, Khan AS, Ksiazek TG, Shobokshi O. Rift Valley fever epidemic in Saudi Arabia: epidemiological, clinical, and laboratory characteristics. *Clin Infect Dis.* 2003 ; 37 : 1084-92.

Mancao MY, Law IM, Roberson-Trammell K. California encephalitis in Alabama. *South Med J.* 1996; 89: 992-93.

Mathiot CC, Gonzalez JP, Georges AJ. Current problems of arboviruses in central Africa. *Bull Soc Pathol Exot Filiales.* 1988; 81: 369-401.

Mayo D, Karabatsos N, Scarano FJ, Brennan T, Buck D, Fiorentino T, Mennone J, Trans S. Jamestown Canyon virus: seroprevalence in Connecticut. *Emerg infect Dis*. 2001; 7: 911-12.

McJunkin JE, Khan RR, Tsai TF. California - La Crosse encephalitis. *Infect Dis Clin North Am*. 1998; 12: 83-93.

McLean RG, Ubico SR, Bourne D, Komar N. West Nile virus in livestock and wildlife. *Curr Top Microbiol Immunol*. 2002; 267: 271-308.

Mekonnen S, Bryson NR, Furie LJ, Peter RJ, Spickett AM, Taylor RJ, Strydom T, Horak IG. Acaricide resistance profiles of single – and multi – host ticks from communal and commercial farming areas in the Eastern Cape and North-West Provinces of South Africa. *Onderstepoort J Vet Res*. 2002; 69: 99-105.

Mele P, Marroni M, Di Candilo F, Moretti MV, Stagni G. Importation dengue. *Rec Prog. Med*. 2001; 92: 37-39.

Mellor PS, Leake CJ. Climatic and geographic influences on arboviral infections and vectors. *Rev Sci Tech*. 2000; 19: 41-54.

Messer WB, Gubler DJ, Harris E, Sivananthan K, de Silva AM. Emergence and global spread of a dengue serotype 3, subtype III virus. *Emerg Infect Dis.* 2003; 9: 800-809.

Mitchell CJ, Morris CD, Smith GC, Karabatsos N, Vanlandingham D, Cody E. Arboviruses associated with mosquitoes from nine Florida counties during 1993. *J. Am Mosq Control Assoc*. 1996; 12: 255-62.

Molina OM, Morales MC, Soto ID, Pena JA, Haack RS, Cardozo DP, Cardozo JJ. Venezuelan equine encephalitis. 1995 outbreak: clinical profile of the case with neurologic involvement. *Rev Neurol*. 1999; 29: 296-98.

Monath TP, Cropp CB, Harrison AK. Mode of entry of a neurotropic arbovirus into the central nervous system. Reinvestigation of an old controversy. *Lab Invest*. 1983; 48: 399-410.

Morse RP, Bennish ML, Darras BT. Eastern equine encephalitis presenting with a focal brain lesions. *Pediatr Neurol*. 1992; 8: 473-75.

Nalca A, Fellows PF, Whitehouse CA. Vaccines and animal models for arboviral encephalitides. *Antiviral Res*. 2003; 60: 153-74.

Navarro JM, Fernandez-Roldan C, Perez-Ruiz M, Sanbonmatsu S, de la Rosa M, Sanchez-Seco MP. Meningitis by Toscana virus in Spain: description of 17 cases. *Med Clin (Barc).* 2004; 27: 420-22.

Oelofsen MJ, Van der Ryst E. Could bats act as reservoir hosts for Rift Valley fever virus ? *Onderstepoort J Vet Res*. 1999; 66: 51-54.

Ogen-Odoi A, Miller BR, Happ CM, Maupin GO, Burkot TR. Isolation of Thogoto virus (Orthomyxoviridae) from the banded mongoose, Mongos mungo (Herpestidae), in Uganda. *Am J Trop Med Hyg*. 1999; 60: 439-40.

Okuno T, Oya A, Ito T. The identification of Negishi virus, a presumably new member of Russian spring-summer encephalitis virus family isolated in Japan. *Jpn J Med Sci Biol*. 1961; 14: 51-59.

Patey O, Ollivaud L, Breuil J, Lafaix C. Unusual neurologic manifestations occurring during dengue fever infection. *Am J Trop Med Hyg*. 1993; 48: 793-802.

Patz JA, Epstein PR, Burke TA, Balbus JM. Global climate change and emerging infectious diseases. *JAMA* 1996; 275: 217-223.

Pavri K. Clinical, clinicopathologic, and hematologic features of Kyasanur Forest disease. *Rev Infect Dis.* 1989; 11 (S4): 854-59.

Peterhans E., Zanoni R., Bertoni G. How to succeed as a virus: strategies for dealing with the immune system. *Vet Immunol Immunopathol.* 1999; 72: 111-117.

Pisecky T., Freund C. The Austrian TBE Patient Support Group. *Vaccine* 2003; 21 (S1): 73-74.

Potula R, Badrinath S, Srinivassan S. Japanese encephalitis in an around Pondicherry, South India: a clinical appraisal and prognostic indicators for the outcome. *J Trop Pediatr.* 2003; 49: 48-53.

Powell KE, Kappus KD. Epidemiology of St.Louis encephalitis and other acute encephalitides. *Adv Neurol.* 1978; 19: 197-213.

Przelomski MM, O'Rourke E, Grady GF, Berardi VP, Markley HG. Eastern equine encephalitis in Massachussetts: a report of 16 cases, 1970-1984. *Neurology.* 1988; 38: 736-39.

Rehle TM. Classification, distribution and importance of arboviruses. *Trop Med Parasitol.* 1989; 40: 391-95.

Reisen WK. Epidemiology of St Louis encephalitis virus. *Adv Virus Res.* 2003; 61: 139-83.

Rodhain F. Problems posed by the spread of Aedes albopictus. *Bull Soc Pathol Exot.* 1996; 89: 137-40.

Rodhain F. The idea of natural reservoir in arbovirology. *Bull Soc Pathol Exot.* 1998; 91: 279-82.

Romano-Lieber NS, Iversson LB. Serological survey on arbovirus infection in residents of an ecological reserve. *Rev Saude Publica.* 2000; 34: 236-42.

Romero JR, Newland JG. Viral meningitis and encephalitis: traditional and emerging viral agents. *Semin Pediatr Infect Dis.* 2003; 14: 72-82.

Romi R. Aedes albopictus in Italy: an underestimated health problem. *Ann Ist Super Sanità* 2001; 37: 241-47.

Rust RS., Thompson WH., Matthews CG., Beaty BJ., Chun RW. La Crosse and other forms of California encephalitis: *J Child Neurol.*, 1999; 14: 1-14.

Saxena VK. Ixodid ticks infesting rodents and sheep in diverse biotopes of southern India. *J Parasitol.* 1997; 83: 766-67.

Schlesinger JJ, Chapman S, Nestorowicz A, Rice GM, Ginocchio TE, Chambers TJ. Replication of Yellow fever virus in the mouse central nervous system: comparison of neuroadapted and non-neuroadapted virus and partial sequence analysis of the neuroadapted strain. *J Gen Virol.* 1996 ; 77 : 1277-85.

Schlesinger S. Alphavirus vectors: development and potential therapeutic applications. *Espert Opin Biol Ther.* 2001; 1: 177-91.

Schwarz TF, Jager G, Gilch S, Pauli C, Eisenhut M, Nitschko H, Hegenscheid B. Travel-related vector-borne virus infections in Germany. *Arch Virol Suppl.* 1996; 11: 57-65.

Scott TW. Are bats really involved in dengue virus transmission? *J Med Entomol.* 2001; 38: 771-72.

Shamanin VA, Pletnev AG, Rubin SG, Zlobin VI. The differentiation of viruses of tick-borne encephalitis complex by means of RNA-DNA hybridization. *Vopr Virusol*. 1991; 36: 27-31.

Siam AL, Meegan JM, Gharbawi KF. Rift Valley fever ocular manifestations: observations during the 1977 epidemic in Egypt. *Br J Ophthalmol*. 1980; 64: 366-74.

Smith JL, Fonseca DM. Rapid assay for identification of members of the Culex (Culex) pipiens complex : their hybrids, and other sibling species (Diptera: Culicidae). *Am J Trop. Med Hyg*. 2004; 70: 339-45.

Solomon T, Dung NM, Vaughn DW, Kneen R, Thao LT, Raengsakulrach B, Loan HAT, Day NP, Farrar J, Myint KS, Warrell MJ, James WS, Nisalak A, White NJ. Neurological manifestations of dengue infection. *Lancet*. 2000; 335: 1053-59.

Solomon T, Mallewa M. Dengue and other emerging flaviviruses. *J Infect*. 2001; 42: 104-5.

Spicer PE. Japanese encephalitis in Western Irian Jaya. *J Travel Med*. 1997; 4: 146-47.

Studdert MJ, Azuolas JK, Vasey JR, Hall RA, Ficorilli N, Huang JA. Polymerase chain reaction tests for the identification of Ross River, Kunjin and Murray Valley encephalitis virus infections in horses. *Aust Vet J*. 2003; 81: 76-80.

Tolou H, Durand JP, Pisano MR. Current status of dengue. *Med Trop*. 1997; 57: 70-73.

Tsai TF, Cobb WB, Bolin RA, Gilman NJ, Smith GC, Bailey RE, Poland JD, Doran JJ, Emerson JK, Lampert KJ. Epidemiologic aspects of a St. Louis encephalitis outbreak in Mesa County, Colorado. *Am J Epidemiol*. 1987; 126: 460-73.

Utz JT, Apperson CS, MacCormack JN, Salyers M, Dietz EJ, McPherson JT. Economic and social impacts of La Crosse encephalitis in western North Carolina. *Am J Trop Med Hyg*. 2003; 69: 509-18.

Valassina M, Cusi MG, Valensin PE. A mediterranean arbovirus: the Toscana virus. *J Neurovirol*. 2003; 9: 577-83.

Van Den Hurk AF, Johansen CA, Zborowski P, Paru R, Foley PN, Beebe NW, Mackenzie JS, Ritchie SA. Mosquito host-feeding patterns and implications for Japanese encephalitis virus transmission in northern Australia and Papua New Guinea. *Med Vet Entomol*. 2003; 17: 403-11.

Vanlandingham DL, Davis BS, Lvov DK, Samokhvalov E, Lvov DS, Black WC, Higgs S., Beaty BJ. Molecular characterization of California serogroup viruses isolated in Russia. *Am J Trop Med Hyg*. 2002; 67: 306-9.

Venugopal K, Gritsun T, Lashkevich VA, Gould EA. Analysis of the structural protein gene sequence shows Kyasanur Forest disease virus as a distinct member in the tick-borne encephalitis virus serocomplex. *J Gen Virol*. 1994 ; 75 : 227-32.

Verani P., Ciufolini MG., Nicoletti L. Circulation of TBE virus in Italy: seroepidemiological and ecovirological studies. *International Symposium on TBE, Badenwien, Wien*, October 19-20, 1979.

Verani P., Ciufolini MG., Nicoletti L. Arbovirus surveillance in Italy. *Parassitologia* 1995; 37: 105-8.

Verani P, Nicoletti L, Ciufolini MG. Antigenic and biological characterization of Toscana virus, a new Phlebotomus fever group virus isolated in Italy. *Acta Virol*. 1984; 28: 39-47.

Vernet G. Diagnosis of zoonotic viral encephalitis. *Arch Virol Suppl*. 2004; 18: 231-44.

Walsh JF, Molyneux DH, Birley MH. Deforestation: effects on vector-borne disease. *Parasitology* 1993; 106 (S): 55-75.

Walters LL, Tirrell SJ, Shope RE. Seroepidemiology of California and Bunyamwera serogroup (Bunyaviridae) virus infections in native populations of Alaska. *Am J Torp Med Hyg.* 1999; 60: 806-21.

Widjaja S, Soekotjo W, Hartati S, Jennings GB, Corvin AL. Prevalence of hemagglutination-inhibition and neutralizing antibodies to arboviruses in horses of Java. *Southeast Asian J Trop Med Public Health*. 1995; 26: 109-13.

Williams KH, Hollinger FB, Metzger WR, Hopkins CC, Chamberlain RW. The epidemiology of St. Louis encephalitis in Corpus Christi, Texas, 1966. *Am J Epidemiol.* 1975; 102: 16-24.

Zeller HG. West Nile virus: a migration arbovirus of current interest. *Med Trop.* 1999; 59: 490-94.

Chapter 6

Neurologic Manifestations of West Nile Virus Infection

Ronen Spiegel and Yoseph Horovitz*
Pediatric Department A, HaEmek Medical Center, Afula,
Rappaport School of Medicine, Technion, Haifa

Abstract

West Nile virus (WNV) infection is a zoonotic disease. The virus belongs to the family Flaviviridae. Wild birds serve as the reservoir of the virus in nature and mosquitoes mainly from the culex species serve as the vector for its replication cycle.

Humans as well as horses and other domestic animals are incidental hosts.

Until recently, WNV infections were traditionally considered to be a mild self-limited disease. When symptomatic it usually manifested with fever, headache, myalgia, generalized lymphadenopathy and rash. Central nervous system (CNS) involvement was infrequent. In the last decade, several epidemics of WNV have occurred in several areas in Europe, in Israel and, for the first time, in the Western Hemisphere, accounting for a large outbreak in New York. These epidemics were characterized by an exceptionally high rate of neurological morbidity and mortality.

The most common neurological involvement are encephalitis/meningoencephalitis followed by aseptic meningitis and flaccid paralysis. Other less common neurological presentations include Guillain-Barre syndrome, rhombencephalitis, optic neuritis and aphasia. The cause for this clinical evolution from a mild flu-like illness to a severe neuroinvasive disease is not fully understood and is thought in part to be related to the emergence of more neurovirulent WNV strains.

This article will discuss the wide range of neurologic manifestations of WNV infections, as well as the pathogenesis of WNV nervous system involvement and current and future therapeutic strategies.

---

* Tel: 972-4-6494316, Fax: 972-4-6495251, E-mail: spiegelr@zahav.net.il

West Nile Virus (WNV)

This virus was first isolated from the blood of a febrile woman in the West Nile district of Uganda in 1937 [1]. The virus is a member of the genus Flavivirus, family Flaviviridae which also includes Japanese encephalitis, St. Louis encephalitis, Murray Valley encephalitis and Kunjin viruses [2]. WNV is a small spherical single stranded positive sense RNA virus of approximately 11,000 nucleotides associated with a core protein (nucleocapsid) and wrapped by a viral envelope which contains several proteins (of most importance is the envelope (E) protein), thought to be responsible for tissue tropism, attachment and entry into host cells and stimulation of host immune responses [3].

Phylogenetic studies demonstrated two distinct lineages of the virus. Strains belonging to lineage I have a worldwide distribution and are responsible for all the outbreaks occurring during the last decade including the 1996 Romania, 1999 Rusian (Volvograd), 1999 and 2000 United States and 2000 Israel. Lineage II strains are distributed mostly in Africa [4].

Transmission and Replication

WNV is an arthropod-borne virus (arbovirus) which means it is transmitted between vertebrate hosts by mosquitoes serving as vectors. In nature, the virus has two replication cycles. The first is in mosquitoes, mainly from the culex species, and the second (amplifying) cycle occurs in wild birds particularly crows. Infected mosquitoes carry the virus in their gut and salivary glands and infect susceptible birds by biting them. The infected birds demonstrate a prolonged period and high level viremia enabling an amplification cycle with every new mosquito bite [5]. Humans as well as other domestic animals exhibit a shorter, low level, viremia period preventing the creation of a new transmission cycle and therefore serve as a dead-end hosts.

Despite this axiomatic hypothesis, several novel modes of transmission were reported including vertical infection from mothers to babies through breastfeeding, in utero transplacental infection, via transplanted infected organ or tissue, and percutaneous exposure from infected blood by laboratory and hospital workers [59].

This avian-mosquito-human interplay is considered to be the mode of introduction of this virus to the American continent at the end of the last century, either by migratory birds, imported infected birds or viremic humans. This global expansion of the virus is responsible for the 1999 New York West Nile outbreak [6,7].

Clinical Features of the Non-Neurological West Nile Fever

The incubation period of WNV in humans usually range from 2-6 days, but may extend to 14 days [8]. When clinically symptomatic, West Nile fever is characterized by a flu-like illness beginning with an abrupt onset of high fever that may be accompanied by chills usually lasting 3-5 days. Other non specific symptoms may include general malaise, headache, myalgia, arthralgia, backache, vomiting, nausea, abdominal pain, diarrhea, maculopapular rash, conjunctivitis, and lymphadenopathy [8-10]. The later was thought to be more prevalent in symptomatic patients in previous outbreaks, but is uncommon in recent reports (less than 5%) [11].

Other less common manifestations may present in more severe cases and include hepatosplenomegaly [10], hepatitis, myocarditis [12] and pancreatitis [13].

Abnormal laboratory findings are usually mild and nonspecific and include either mild leukocytosis in less than 50% of the patients or mild leukopenia in less than 10% of the patients [11], mild anemia, mild thrombocytopenia, mild elevation of liver transaminases, and mild hypokalemia [11,14].

Neurologic Manifestations

West Nile (WN) infection predominantly affects the nervous system. Involvement of the central nervous system (CNS) is associated with significant morbidity and mortality rates. In earlier WN epidemics neurologic involvement was considered to be a rare complication and therefore excellent prognosis was the rule in general [8-10]. By contrast, in recent epidemics neurologic manifestations predominated especially in the hospitalized cases [11,14,15].

In general, neurologic symptomatology may be divided into the following three main groups depending on the anatomic location of the nervous system part damaged:

1. Aseptic meningitis – when involving mainly the meninges.
2. Encephalitis or meningoencephalitis – when most of the inflammatory process involves the brain parenchyma with or without the meninges.
3. Myelitis – characterized by predominant spinal cord inflammation.

In some cases two or even three of these manifestations may coexist in the same patient.

Our discussion will concentrate on these three major neurologic presentations. We will also describe in brief other less common central nervous manifestations reported in the literature.

Encephalitis

In several recent wide range WN epidemics, encephalitis was the most common presentation among hospitalized patients accounting for about 60% of hospital admissions [11,14,15]. Still, results of a household based seroepidemiological survey estimated that the 1999 New York epidemic consisted of about 8200 WNV human infections and since there were 59 cases of WN encephalitis (WNE), less than 1% of WNV infections suffered severe neurological disease [16]. The clinical course of WNE is usually abrupt with the triad of fever, headach and altered level of consciousness. This may resemble the clinical features of other causes of viral encephalitis including arthropod-borne viruses, enteroviruses etc. This similar nonspecific features makes it difficult to identify the offending pathogen in viral encephalitis based on clinical base [17]. Seizures were reported to occur more frequently in early reports of WN encephalitis but were uncommon in later outbreaks [11,14]. The cause for this change is not completely understood. The lack of convulsions is in contrast with Japanese encephalitis infections where seizures play a major role in the presentation of the disease [18].

Ataxia and extrapyramidal signs were reported in 17% of the cases in one report [15] but were uncommon in others [11,14]. Cranial nerve abnormalities were relatively rare and when they did occur were associated with high mortality rate [19].

The laboratory findings typical for WN encephalitis include cerebrospinal fluid (CSF) pleocytosis (mean 38-308 cells/mm³, range 0-1750 cells/³) with mild lymphocytic predominance, elevated CSF protein concentration (mean 85-110 mg/dL, range 18-1900 mg/dL) and normal glucose concentration. This CSF indices are nonspecific to the etiologic pathogen of the encephalitis and may be found in most other causes of viral encephalitis [17]. As a rule, when tested, CSF immunoglobulin M (IgM) -capture enzyme immunoassay (EIA) to WNV are usually found during the first week of illness. In the New York/New Jersey 2000 epidemic, almost all patients (18 out of 19) had positive CSF IgM-capture EIA and 9 of them had >four-fold serial change in plaque-reduction neutralizing antibody titer (PRNT) to WNV in paired, appropriately timed CSF samples [19]. By definition demonstration of IgM antibodies to WNV in the CSF by EIA or by >four-fold rise in PRNT antibodies are diagnostic for recent WNV infection with CNS involvement [20]. Detection of WNV in the CSF by either PCR techniques or by virus isolation is diagnostic for current CNS infection, however its utility in clinical practice is of limited value. In a study of patients serologically diagnosed as WNV meningoencephalitis using TaqMan reversed transcript (RT)-PCR on CSF samples the sensitivity was only 57% [21]. The sensitivity of RT-PCR for detection of WNV nucleic acids from the CSF was even lower during the New York/New Jersey 2000 epidemic as only 1 out of 13 confirmed meningoencephalitis patients was tested positive making this assay non reliable in daily non research practice [19].

Electroencephalogram (EEG) studies performed on encephalitic patients usually reveal generalized slow waves without convulsive tendency consistent with nonspecific encephalitis. No specific pattern for WNV infection was reported [11].

Brain computerized tomography (CT) performed in the acute encephalitic phase is normal in most cases and may reveal preexisting chronic lesions such as old infarcts, cortical atrophy and old hemorrhages and therefore this imaging technique is of no diagnostic value in these patients [11,14,19]. Initial brain magnetic resonance imaging (MRI) revealed nonspecific leptomeningeal and periventricular enhancement in 30% of the patients tested consistent with acute inflammatory process. However, this findings are non specific for WN infection [14].

Histologic features of WN encephalitis as demonstrated from autopsies of four patients in the 1999 New York epidemic showed minimal non specific evidence of viral inflammation in brain tissue. This included microglial nodules, perivascular cuffing, neuronal degeneration, neurophagia and focal inflammation involving mainly the medulla and the thalamus and less frequently the cerebellum and cerebrum. WNV was identified in autopsy brain tissue by either PCR and immunohistochemical stain [22,23].

In a search for risk factors associated with the development of WNE, the only independent risk factor found was an advanced age. This age related susceptibility was clearly demonstrated by the correlation of patients age groups versus symptomatic infection in general and versus CNS involvement in specific [15]. This unique age distribution is completely opposite to the age distribution of Japanese encephalitis virus infections where most of the CNS morbidity occurs among the pediatric group sparing the elderly group [18]. Although WN infection in the pediatric group usually carries a mild course with low CNS morbidity, children with immune deficiency states such as patients receiving chemotherapy, are at increased risk for severe CNS WN infection [24]. Immunocompromised patients were more vulnerable to severe WNV infection in all age groups. We found two case reports of HIV infected patients with CD4 counts below 200 cells/mm^3 who developed severe encephalitis [60]. Other case reports of immune deficient patients included two patients with lymphoma who developed WN encephalitis while receiving chemotherapy treatment. In both patients reported, the encephalitis evolved during the neutropenic phase [24,59]. This apparent relation was reinforced by the findings found during the 2000 Israeli outbreak where immunocompromised patients experienced statistically significant higher mortality rates compared with immunocompetent patients (31% vs. 13%) [11].

Apart from old age and the probable consequential immune deficiency, no underlying medical condition such as hypertension, chronic lung disease, or diabetes mellitus was statistically correlated with the development of encephalitis. However, encephalitic patients were found on a case control study to spend more time outdoors daily during the epidemic period. The authors speculated that this association is related to exposure to more bites of infected mosquitoes which increase the viral inoculum sufficient to invade the neural system to create a clinically significant infection [25].

The outcome of WN encephalitis is not favorable. As a rule all the fatalities reported were patients initially presenting with encephalitis that had a further clinical deterioration. Overall mortality/case ratio ranged from 4.3% [14,15] to 14% [11]. Among encephalitic patients alone the fatality rate was even higher reaching 19% in New York 1999 epidemic and 24% in the Israeli 2000 epidemic [11,14]. Among encephalitic patients over 75 years the prognosis was the poorest [14]. Recovery rates of WNE from one epidemic found that 37% recovered completely and 53% recovered partially 26% were ambulatory with assistance and 11% were bedridden. 26% required outpatient physical therapy, 16% required speech therapy, and 11% required occupational therapy after hospital discharge [19].

Aseptic Meningitis

WN aseptic meningitis is clinically similar to all other etiologic viral meningitis [26]. Its clinical features include fever, headache, vomiting and signs of meningeal irritation (stiff neck, Brudjinski sign, Kernig sign). The lack of mental status change differentiates it from WN meningoencephalitis.

During the recent WN epidemics aseptic meningitis comprised 16%, 29% and 42% of all documented WN cases [11,14,19]. The clinical course is milder than that of encephalitis/meningoencephalitis and complete recovery without neurologic sequels should be expected. Typical CSF indices include pleocytosis with lymphocytic predominance, although some reports described a polymorphonuclear predominance most probably due to the early stage of the disease the lumbar puncture was performed [27], elevated protein concentration, normal glucose concentration and negative culture results.

In general patients presenting with aseptic meningitis tend to be younger than patients presenting with encephalitis. This age related clinical difference proved to be statistically significant during the Israeli 2000 epidemic [11].

Acute Myelitis

Encephalomyelitis epidemics were reported previously in horses and birds following an infection with WNV. These reports described the concomitant expression of meningoencephalitis and acute myelitis (polio-like and non polio-like syndromes) in these animals [28-30].

The first report of acute anterior myelitis complicating WN fever in humans appeared in 1979. In their report Gadoth et al described a 22-year-old Israeli scuba diver who presented with an acute febrile illness and developed flaccid paresis of the lower left extremity with loss of deep tendon reflexes but without mental changes or sensory changes. Initial lumbar puncture revealed 94 cells/mm^3 and elevated protein concentration of 63 mg/dL. Subsequent spinal taps showed decrease in the WBC count and increase in the protein concentration. His disease natural course was of gradual recovery with resolution of muscle strength and deep tendon reflexes but mild diminution of the muscle strength of the affected limb remained [31].

Since this report no other WN cases of acute myelitis were documented until the epidemics in Romania and Russia during the last decade of the former century. In these reports 15-20% of the patients demonstrated variable degree of muscle weakness as well as lower extremities paresis or paralysis as part of their WN infection [32,33]. During the recent wide epidemics of WN in the United States and in Israel muscular and peripheral nervous involvement became a prominent neurological manifestation presenting as either non specific muscle weakness or flaccid paralysis. The New York 1999 epidemic reached the highest rate in this respect when according to Nash et al 27% of the patients experienced decreased muscle strength, 32% of the patients had hyporeflexia or afeflexia, and 10% of the patients had "true" flaccid paralysis without ascending or descending pattern of progression [14]. The clinical variability is wide and may range from muscle weakness alone involving the lower limbs without tendon reflex changes on the one hand, to quadriplegia with urinary retention due to bladder dysfunction, dysphagia and respiratory distress necessitating mechanical ventilation, on the other hand. The paresis may be symmetric or asymmetric and may involve either the upper limbs, lower limbs or both. Sensory function is usually spared. Laboratory findings in WN myelitis are similar to those found in central nervous WN infection (encephalitis/aseptic meningitis). When obtained the CSF reveals mild to moderate pleocytosis with lymphocyte predominance, elevated protein concentration, normal glucose concentration and negative culture results. When subsequent spinal taps are performed the WBC count tend to decrease concomitantly with increase in the protein concentration

[14,22]. When obtained WNV specific IgM capture antibodies were detected in the CSF by EIA [14,22].

MRI of the complete spinal cord was normal in the few patients it was performed and therefore is usually unnecessary if the diagnosis of WNV myelitis is definite [34,35]. Electromyography (EMG) and nerve conduction velocity (NCV) are the laboratory procedures most informative in these cases. In the New York 1999 epidemic 80% of the patients who underwent EMG and NCV studies demonstrated abnormal findings. This included reduced compound muscle potentials and fibrillation potentials, and reduced motor amplitudes with either normal or slightly decreased nerve conduction velocities[14,22]. These results were consistent with axonal type polyneuropathy with sparing of the sensory fibers and differed from the findings usually found in Guillain-Barre syndrome where nerve conduction studies usually reveal prolonged motor and sensory conduction rates due to demyelination [36].

In the last year two brief reports were published describing a specific poliomyelitis-like syndrome typical to WN infection in four different patients [37,38]. Polio virus infection produces a well-defined specific clinical entity characterized by the triad of asymmetric flaccid paralysis, diminished or absent tendon reflexes of the affected limbs and no paresthesias or sensory loss. This clinical syndrome is the result of the virus ability to attach to receptors in the anterior horn of the spinal cord. Laboratory findings typical to polio infection include CSF pleocytosis and elevated protein concentration [39]. Since the almost global eradication of polio virus from the developed countries with the aid of the combined inactivated polio vaccine (IPV) and the live attenuated oral polio vaccine (OPV), paralytic polio-like syndrome are caused by other RNA viruses including enteroviruses, echoviruses and coxsackiviruses [40]. Several case reports have associated poliomyelitis-like syndrome to other flaviviruses including Japanese encephalitis virus and dengue fever [41,42]. In 2003 Sejvar et al reported the greatest group of patients ever published with poliomyelitis-like syndrome due to WN infection. This report included a detailed clinical description, laboratory, and electrodiagnostic workup of seven patients who underwent a thorough evaluation. All patients present with acute asymmetric flaccid paralysis, decreased or absent tendon reflexes in the affected limb, and typical electrophysiologic findings suggesting anterior horn damage without sensory abnormalities [34].

All the patients were from Mississippi and Louisiana and were ill within a two-month period (July-August 2002). The age range of the patients was 39 to 69 years (mean 53 years old). This age group is significantly younger than the age group of WNE. Three of the seven patients reported had nuchal rigidity consistent with meningitis and three of the seven had mental status changes consistent with encephalitis. 5 of the patients had asymmetric upper limb involvement and 3 had asymmetric lower limb involvement. Electrodiagnostic studies showed reduced compound muscle action potentials of the affected extremity, widespread denervation and normal sensory nerve action potentials. These findings are consistent with severe asymmetric process affecting the anterior horn. Myopathy, demyelinating polyneuropathy and diffuse axonal polyneuropathy were not evident [34,43].

We found only one report of autopsy examination of the spinal cord in a patient with WN meningoencephalitis and polio-like paralysis. The macroscopic examination of the spinal cord was non remarkable. The microscopic examination showed a non specific chronic inflammatory infiltrate throughout the length of the spinal cord more prominent in the lumbar region [44].

The prognosis of WN myelitis depends on several factors. Bad prognostic factors include the presence of encephalitis, elderly patients, and severe spinal cord dysfunction. Death was either the result of severe encephalitis or the consequence of respiratory failure due to paralysis and malfunction of the respiratory muscles.

Mild cases (only muscle weakness) usually achieved complete recovery but in most cases of moderate to severe paralysis certain degree of weakness remained [14,19,34,36]. In the subgroup of polio-like syndrome most of the patients remained with a certain degree of limb weakness [37,38].

In general WN encephalitic patients accompanied by muscle weakness were more likely to bear a poorer prognosis compared with encephalitic patients with preserved muscle strength. In the first group the case fatality rate reached 30% [14].

In summary, WNV infection is associated with a heterogenous neurologic manifestations that include muscle weakness, flaccid paralysis and a well defined poliomyelitis-like syndrome. A strict detailed clinical documentation along with laboratory tests, electrophysiologic studies and neuroimaging should be conducted in order to better define and understand the etiology and pathogenesis of this WN flaccid paralysis clinical spectrum.

Rare Neurological Manifestations

Aside from the three main neurologic manifestations of WNV infection i.e encephalitis, aseptic meningitis and acute myelitis (poliomelitis-like and non poliomyelitis-like) other rare neurologic manifestations were reported in the recent epidemics mainly as isolated case reports.

Guillain-Barre syndrome (GBS) – During the 1999 New York WN epidemic Ahmed et al described a 65-year-old man who presented with decreased muscle strength of upper and lower limbs associated with areflexia numbness and sensory changes in his hands and feet. During his hospital stay he developed bilateral peripheral facial nerve palsy, his weakness progressed and he became bedridden and as a result of a further respiratory distress he required mechanical ventilation. CSF sample revealed elevated protein concentration without WBC. EMG and NCV studies revealed prolonged distal motor and sensory latencies, reduced compound motor action potentials (CMAP), and unobtainable F and H waves. These findings were consistent with demyelinating polyneuropathy with secondary motor axon degeneration. WNV-PCR and WNV specific IgM-capture antibodies from the CSF were both positive. The patient was diagnosed as GBS secondary to WN infection. Despite 5 courses of plasmapheresis and two courses of intravenous gammaglobulins he experienced only slight improvement of muscle strength [45].

GBS is an autoimmune disorder of the peripheral nervous system resulting in inflammation and demyelination hence it is often termed acute demyelinating polyradiculoneuropathy (AIDP). GBS patients typically present with symmetric muscle weakness initially affecting the lower limbs and progressing in an ascending form to involve upper limbs and respiratory muscles [46]. Classically, abnormal sensation such as tingling and numbness in the hands and face accompany the motor deficits. Tendon reflexes are either reduced or absent. The CSF protein concentration is elevated with a paucity of WBC. This phenomenon was termed albuminocytologic dissociation. Several infectious pathogens are associated with preceding the development of the syndrome by a mechanism of molecular mimicry in which the immune response to specific antigens or infectious organisms attack similar epitopes in the peripheral nervous system of the host. In this respect the most common pathogens known to cause GBS are cytomegalo virus (CMV), Epstein-Barr virus (EBV), campylobacter jejuni, mycoplasma pneumonia, several herpes viruses, etc [36]. The precise mechanism in which WNV causes GBS is yet to be discovered. We assume that WNV induced GBS is a unique clearly defined presentation that might be included within the heterogenous group of WN myelitis.

Optic neuritis – During the Israeli 2000 WN outbreak Vaispapir et al. described a 55-year-old woman who was admitted with clinical symptoms symptoms consistent with meningoencephalitis. Her CSF contained 400 WBC/mm^3, elevated protein concentration, normal glucose concentration and a negative culture. During her hospitalization she experienced neurologic deterioration with the development of right abducens and facial cranial nerve palsies, blurred vision with progressive worsening of visual acuity and severe cerebellar ataxia. Ophthalmologic examination revealed bilateral optic disk swelling more prominent on the right side, constriction of visual field, and relative afferent pupillary defect (ARPD). Diagnosis of acute WNV infection was made by positive serum specific WNV-IgM antibodies. She gradually recovered on conservative treatment and was discharged after 30 days. Mild ataxia, blurred vision and right optic disc swelling were still present on discharge and long-term follow up was not reported [47].

Optic neuritis is a self limited condition characterized by unilateral or bilateral blurring of vision which may progress to almost complete loss of vision over days to weeks. Other clinical features include pain in the affected eye, loss of colour vision, visual field defects, and RAPD on ophthalmologic examination [48]. The most common underlying cause of optic neuritis is multiple sclerosis. Viral infections may cause optic neuritis by two mechanisms. This includes viral induced optic neuritis as part of a CNS infection most commonly by neurovirulent pathogens. The second mechanism is a postinfectious inflammatory optic neuritis. In the latter, the patient usually presents with bilateral and simultaneous eye involvement. Based on the case report described above, it appears that WNV may cause optic neuritis by direct inflammation of the optic nerve as well as other cranial nerves (abducens and facial nerves).

Although systemic steroids may hasten the speed of visual recovery it does not improve the ultimate visual outcome [48]. Therefore in WNV associated optic neuritis there seems to be no additional role for steroid treatment at present and expectant approach should be preferred by the physician in charge. One should also weigh the potential risk of exacerbating the disease course by using steroids.

Rhombencephalitis – During the 1999 New York outbreak Nichter et al described a 15-year-old boy who was hospitalized because of fever, headache, and acute confusion. The CSF examination on admission revealed pleocytocis with lymphocytic predominance and elevated protein concentration. During his stay he developed difficulties in swallowing due to depressed gag reflex which necessitated the insertion of nasogastric tube for feeding, truncal ataxia and bilateral facial weakness. The boy experienced a gradual course of recovery, with the use of auxiliary supportive treatment, which lasted six months. Positive CSF specific IgM antibodies to WNV confirmed the etiologic cause of his disease [49].

Rhombencephalitis is the term used to define specific encephalitis involving the brain stem. In most cases the cause of rhomencephalitis is infectious. The most common etiologic pathogens are: Listeria monocytogenes, Micoplasme pneumonia, herpes simplex viruses, EBV, influenza A virus, varicella, adenovirus, echoviruses and enteroviruses [50-52]. Perhaps, the most important etiologic pathogen associated with this relatively rare complication is enterovirus 71. In 1998 a wide epidemic of this virus was reported in Taiwan, accompanied by a high rate of morbidity and mortality due to rhombencephalitis [40]. Among the flaviviruses only Japanese encephalitis virus was reported previously to cause rhombencephalitis although this manifestation is rare in the clinical spectrum of this virus [18].

MRI studies performed in patients with rhombencephalitis showed that the majority of lesions are located in the pontine tegmentum followed by the medulla oblongata, midbrain, and dentate nuclei in decreased order [40,53]. The prognosis depends on the severity of rhombencepalitis. Grade I lesions are associated with cranial nerve palsies and ataxia and have a milder self limited course in most cases. Grade III lesions are associated by cardiopulmonary failure and significant morbidity and mortality [40].

Unfortunately, in the WN patient described by Nichter MRI was not performed during the acute stage of the illness. Late image during the convalescent stage was unremarkable [49].

Aphasia – During the 2000 Israeli WN epidemic we reported a 4-year-old boy with Hodgkin's lymphoma receiving chemotherapy treatment who was admitted with acute febrile illness. During his stay he became neutropenic in parallel to the development of meningoencephalitis. His serum WNV-IgM specific antibodies were positive. His disease course was predominated by generalized seizures and global aphasia. He was treated with oral ribavirin and intensive supportive care including speech therapy. Complete recovery was achieved within months with motor aphasia being the last to recover [24].

Aphasia is a defined neurological entity characterized by the loss of comprehension and formulation of language. Aphasia is usually the result of lesions located in the left hemisphere of the brain. The most common insults known to cause aphasia are stroke, head injury, cerebral tumors, degenerative CNS diseases and CNS infections. There are several types of aphasia defined by their clinical typical features and the exact anatomic location of the offending cerebral lesion. The classic aphasia variants include Broca's aphasia, Wernicke's aphasia, conduction aphasia, global aphasia and transcortical aphasia [54,55].

The child we described presented with features compatible most with the global aphasia variant. The cerebral part known to be affected in global aphasia are the frontal and temporoparietal left hemisphere [55]. Unfortunately, MRI study was not conducted in this patient and therefore there was no imaging correlation with the clinical symptoms.

Pathogenesis of WNV Neural Damage

Following an infected mosquito bite an initial replication develops at the bite site (skin and regional lymph nodes) which produces the primary viremia. The virus then seeds the reticuloendothelial system (RES) where it replicates to produce the second viremia. During this viremia the virus may be isolated from the blood (usually within two days before the beginning of clinical symptoms to four days during the disease course). Isolation of the virus depends on the duration and the level of the viremia with the latter being more important [56,57]. Following the second viremia the virus may reach any organ including the nervous system. Virus replication occurs in monocytes/macrophages and in peripheral tissues mainly the RES. At this point the course of the disease, its severity and the target organs involved depends on both host and viral factors. Host factors include the immune system (humoral and cell mediated), the efficacy of the blood brain barrier (BBB) in preventing invasion to the CNS and other factors such as age, medications, pre morbid state etc. Viral factors include tissue tropism (for example WNV has nervous system tropism), virulence elements, and its ability to change its nucleic acid composition. Perhaps the most important factor in WNV virulence is the E protein which mediates cell attachment and neuroinvasiveness [56]. Finally, intracellular viral-host cell interactions play a major role in the ability of WNV to replicate and to successfully assemble virion particles to exit the infected host cell [57]. Recent study in this field demonstrated in inbred mice an allele termed *Oas 1b* that confers resistance to flaviviruses in mice. A truncated protein product due to a stop codon confers these mutated mice susceptible to flavivirus infection [58]. Expression of the resistant allele of *Oas1b* in susceptible embryo fibroblasts resulted in partial inhibition of the replication of a flavivirus [58]. There is no homologous *Oas 1b* mouse gene in the human genome but this knowledge may play a role in future antiviral medication.

WNV is a neurotropic virus. Evidence of its presence in neural tissues was demonstrated by several techniques. Only recently, the virus was successfully isolated for the first time from the CSF of a patient with WNE [59]. Former molecular studies in patients with WNE detected viral nucleic acids in the CSF using TaqMan RT-PCR technique [14]. Finally, autopsy studies in patients who died from WNE demonstrated by using immunohistochemical staining the presence of viral antigens in neurons, neuronal processes, and areas of necrosis [22,24]. In order to cause nervous system infection WNV should cross the BBB. The port of entry into the CNS is probably via hematogenous spread. The virus may produce a third replication cycle in the endothelial cells at the BBB. Another proposed mechanism of invading the nervous system is by neuronal transport via axons and nerve processes.

By knowing the factors that influence viral entry into the CNS, we are now able to undestand more precisely some of the risk factors related to severe neurologic presentations:

1. Immune deficiency states –We mentioned earlier the high rate of WN encephalitis among patients with certain degree of immunosupression such as HIV [60], malignant diseases with chemotherapy induced myelosupression [24,59], etc. Earlier reports emphasized that in immunocompromised patients there is higher magnitude and duration of viremia [61]. This may promote the virus ability to invade the BBB.

2. Elderly – The association between older age and higher rates of hypertension and vascular diseases such as diabetes mellitus is well documented. These vascular, age related lesions may disrupt cerebral endothelium enabling higher proportion of viral entry into the CNS.

Pathologic studies in WN patients with brain parenchimal disease (encephalitis) and spinal cord disease (myelitis) demonstrated two mechanisms of neural tissue damage. The presence of viral antigens within damaged neuronal and glial cells suggests direct viral induced cytolysis. The second mechanism is an inflammatory neuronal damage demonstrated by perivascular cuffing, microglial nodules, and leukocyte infiltration. Microglial nodules are composed of lymphocytes and histiocytes and represent host inflammatory reaction to neuronal damage [22,24].

Parenchimal damage was more prominent in the brain stem, medulla cerebrum especially the temporal lobes and the proximal spinal cord especially the anterior horn [22,24]. This anatomic predilection of WNV in humans is in correlation with the clinical manifestations described above and may suggest targeted neurotropism. The factors that influence this targeted neurotropism are presently unknown. Further studies are needed to reveal them.

WNV infection in humans elicit both humoral and cell mediated immune response. WNV specific neutralizing antibodies are produced against viral epitopes especially against the E protein which provides viral attachment properties. These antibodies have a significant role in host defense during the acute infection and may be found in the serum of infected patients during the first week of infection (IgM antibodies). The switch to an IgG antibody response occurs after 4-5 days on average [62]. The presence of WNV-IgM antibodies in the CSF reflects CNS involvement since IgM antibodies are unable to cross the BBB. Another evidence for the autonomic WN antibody local production in the CNS was suggested by Tardei et al. These researchers showed that in several WNE patients when recurrent paired serum and CSF samples were taken simultaneously during disease course IgM antibodies appeared in the CSF before they were found in the serum [62]. CSF WNV specific antibodies protect brain parenchyma by reducing viral replication and by interfering with viral attachment to neuronal cells surface. It is believed that after recovery WNV IgG specific antibodies confer long-term immunity against reinfection [63].

We learn about the importance of cell-mediated immunity in host defense against WNV infections from indirect data. For example the higher prevalence of CNS involvement in patients with cell-mediated immune deficiency states such as HIV patients [60], and patients receiving chemotherapy [24,59]. Yet, its exact role and the complex interplay it has with humoral immunity is not understood and further studies are needed to disclose them.

Animal histologic studies in monkeys and hamsters confirmed persistent CNS infection [64,65]. Similar studies were not performed in humans and, therefore, it is not known whether the mechanism for permanent neurological damage is persistent inflammatory process in the brain parenchyma, or irreversible neurologic damage that was caused during the acute infection. We believe, clinical complete recovery after WNE or WN myelitis is correlated with compatible histologic complete resolution.

Treatment of Neural WN Infection

In non neurologic WN infections the treatment is symptomatic. Self limited course is expected in most uncomplicated cases [56].

Treatment in neurologic WN cases is divided into general supportive treatment and specific antiviral therapy. Supportive care includes anticonvulsive medications in cases where seizure activity predominates the clinical course. The most frequent cause of death in WNE is cerebral edema [66]. Early diagnosis of this life-threatening condition should be followed by immediate initiation of mannitol. Steroids also may be used to reduce cerebral edema although their beneficial use should be carefully weighed against their potential risk of exacerbating the viral infection. WNE may require intensive care setup in severe complicated cases including mechanical ventilation, nasogastic feeding etc. During the recovery stage patients may benefit from physical therapy, occupational therapy, and speech therapy depending on their degree of disability [19,66].

There is no established antiviral treatment for WNE or other flavivirus infection. Several antiviral agents were studied in-vitro and in vivo. For most agents we found only anecdotal case reports. Double blind placebo controlled studies are therefore needed to confirm efficacy of all the proposed medications.

Antiviral therapy for WN encephalitis may be divided into three categories: 1. Purine and pyramidine analogues (ribavirin); 2. Interferon α 2b; and 3. Intravenous immunoglobulins.

Ribavirin is a synthetic guanosine analogue that inhibits replication of DNA and RNA viruses [67]. Ribavirin is effective against hepatitis C virus which also belongs to the flaviviridae family. This evidence raised the hypothesis that it also may be effective against other flaviviruses such as WNV, Japanese encephalitis virus and St Louis encephalitis virus. In vitro studies on incubated human neural cells demonstrated that high doses of ribavirin are effective in inhibiting WNV replication as measured by RT-PCR, and in decreasing the cytopathic effect of WNV on neural cells [68] Based on these data several case descriptions of ribavirin treatment in WNE were reported with variable degrees of success [11,24,69]. Controlled studies are still needed to confirm its efficacy. Wide range epidemics such as the 1996 Romanian, 1999 New York, and 2000 Israeli may serve as population base for this kind of study, but strict inclusion criteria should be defined ahead. Another nucleoside analogue pyridazine nucleoside was found to be effective in preventing WN replication in in-vitro studies.

Interferon α 2b is a known antiviral agent in routine use for hepatitis C infection. Open clinical trials using this agent in other flaviviruses (St. Louis encephalitis and Japanese encephalitis viruses) have shown promising results thus far [70,71]. Several case reports of the use of interferon alpha in WN patients were reported with variable degree of success [69]. Based on these data an open randomized clinical trial of the efficacy of interferon alpha in WN patients is conducted by Rahal. Eligible criteria for entering the study include age above 50 years and initiation of therapy within the first four days of hospitalization (in order to prevent irreversible CNS damage in this high risk age group) [72].

Intravenous immunoglobulin preparation was shown to be dramatically effective in an Israeli 70-year-old WNE patient who was in a coma on initiation of treatment. When checked for the content of specific WNV antibodies this preparation was found to have a high content of the specific antibodies. This is most probably a reflection of the endemic character of this virus in Israel [73]. The protective role of specific WNV antibodies in host immunity was comprehensively discussed above. We conclude that specific WNV immunoglobulin preparations may be useful in severe cases.

Regardless of antiviral agent used, our aim is to begin treatment early in the course of the disease in order to prevent the development of encephalitis. To be able to initiate early therapy one must have rapid, widely available, and highly reliable diagnostic tests.

Finally, at present no specific WNV immunization is available although research in this field is active [74,75]. The development of effective formalin inactivated vaccine for Japanese encephalitis virus [17] is encouraging that similar vaccination for WNV will be available in the future.

In the absence of such reliable vaccine prevention of WNV, infections will continue to depend on public education, national vector (mosquito) control programs, and prospective surveillance of infected mosquitoes, birds and horses [76-78].

References

[1] Smithburn KC, Hughes TP, Burke AW, Paul JH. A neurotropic virus isolated from the blood of a native Uganda. *Am. J. Trop. Med.* 1940; 20:471-492.

[2] Gubler DJ, Roehring JT. Togaviruses and Flaviviridae. In: CollierL, Balows A, Sussman M, editors. *Topley and Wilson's microbiology and microbial infections.* London: Arnold Publishing; 1999 p. 579-600.

[3] Heinz FX, Collett MS, Purcell RH, Gould EA, Howard CR, Houghton M et al. Family: Flaviviridae. In: Van Regenmortel MHV, Fauquet CM, Bishop DHL,Carstens EB, Estes MK, Lemon SM et al, editors. Virus taxonomy: classification and nomenclature of viruses. 7th *Report of the International Committee on Taxonomy of Viruses.* San Diego: Academy Press; 1999. p. 859-878.

[4] Lanciotti RS, Ebel GD, Deubel V, et al. Complete genome sequences and phylogenetic analysis of West Nile virus strains isolated from United States, Europe, and the Middle East. *Virology* 2002; 298: 96-105.

[5] Rappole JH, Derrickson SR, Hubalek Z. Migratory birds and spread of West Nile virus in Western Hemisphere. *Emerg. Infect. Dis.* 2000; 6:319-328.

[6] Turell MJ, Sardelis MR, Dohm DJ, O'guinn ML. Potential North American vectors of West Nile virus. *Ann.. NY Acad. Sci.* 2001; 951:317-324.

[7] Lanciotti RS, Roering JT, Deubel V, et al. Origin of West Nile virus responsible for an outbreak of encephalitis in the northeastern United States. *Science* 1999; 286:2333-2337.

[8] Goldblum N, Sterk VM, Paderski B. The clinical features of the disease and theisolation of West Nile virus from the blood nine human cases. *Am. J. Hygiene* 1954; 59:89-103.

[9] Spingland I, Jasinska-Klingberg W, Hofshi E, Goldblum N. Clinical and laboratory observations in the outbreak of West Nile fever in Israel in 1957. *Harefuah* 1958; 54:275-281.

[10] Marberg K, Goldblum K, Sterk VV, Jasinska-Klingberg W, Klingberg MA. The natural history of West Nile fever. I. Clinical observations during an epidemic in Israel. *Am. J. Hygiene* 1956; 64:259-269.

[11] Chowers MY, Lang R, Nassar F, et al. Clinical characteristics of the West Nile fever outbreak, Israel, 2000. *Emerg. Infect. Dis.* 2001; 7:675-678.

[12] Solomon T, Vaughn DW, Clinical features and pathophysiology of Japanese encephalitis and West Nile virus infections. In: Mackenzie JS, Barrett AD, Deubel V, eds. Current topics in microbiology and immunology: *Japanese encephalitis and West Nile virus infections*. Berlin: Springer-Verlag, 2002:171-194.

[13] Perelman A, Stern J. Acute pancreatitis in West Nile fever. *Am. J. Trop. Med. Hyg.* 1974; 23:1150-1152.

[14] Nash D, Mostashari F, Fine A, et al. The outbreak of West Nile infection in the New York city area in 1999. *N. Eng. J. Med.* 2001; 344:1807-1814.

[15] Tsai TF, Popovici F, Cernescu C, Campbell GL, Nedelcu NI. West Nile encephalitis epidemic in southeastern Romania. *Lancet* 1998; 352:767-771.

[16] Mostashari F, Bunning ML, Kitsutani PT, et al. Epidemic West Nile encephalitis, New York, 1999: results of a household-based seroepidemiological survey. *Lancet* 2001; 358:261-264.

[17] Whitlley RJ, Gnann JW. Viral encephalitis: familiar infections and emerging pathogens. *Lancet* 2002; 359:507-514.

[18] Solomon T, Dung NM, Kneen R, Gainsborough M, Vaughn DW, Khanh VT. Japanese encephalitis. *J. Neurol. Neurosurg. Psychiatry* 2000; 68:405-415.

[19] Weiss D, Carr D, Kellachan J et al. Clinical findings of West Nile virus infection in hospitalized patients, New York and New Jersey, 2000. *Emerg. Infect. Dis.* 2001; 7:654-8.

[20] Cennters for Disease Control and Prevention. Case definitions for infectious conditions under public health surveillance. *MMWR Morb. Mortal. Wkly. Rep.* 1997; 46:1-55.

[21] Lanciotti RS, Kerst AJ, Nasci RS, et al. Rapid detection of West Nile virus from human clinical specimens, field collected mosquitoes, andd avian samples by a Taqman reverse transcriptase-PCR assay. *J. Clin. Microbiol.* 2000; 38:4066-4071.

[22] Asnis DS, Conetta R, Teixeira AA, Waldman G, Sampson BA. The West Nile virus outbreak of 1999 in New York: The Flushing hospital experience. *Clin. Infect. Dis.* 2000; 30:413-418.

[23] Shieh WJ, Guarner J, Layton M et al. The role of pathology in an investigation of an outbreak of West Nile encephalitis in New York, 1999. *Emerg. Infect. Dis.* 2000; 6:370-372.

[24] Spiegel R, Miron D, Gavriel H, Horovitz Y. West Nile meningoencephalitis complicated by motor aphasia in hodgkin's lymphoma. *Arch. Dis. Child* 2002; 86:441-442.

[25] Han LL, Popovici F, Alexander JP, et al. Risk factors for West Nile virus infection and meningoencephalitis, Romania, 1996. *J. Infect. Dis.* 1999; 179:230-233.

[26] Rotbart H. Viral meningitis and the aseptic meningitis syndrome. In: Scheld W, Whitley R, Durack D, editors. Infections of the central nervous system. Philadelphia: *Lippincot-Raven Publishers*; 1997. p. 239-263.

[27] Flatau E, Kohn D, Daher O, Varsano N. West Nile fever encephalitis. *Isr. J. Med. Sci.* 1981; 17:1057-1059.

[28] Cantile C, Di Guardo G, Eleni C, Arispici M. Clinical and neuropathological features of West Nile virus equine encephalomyelitis in Italy. *Equine Vet. J.* 2000; 32:31-35.

[29] Cantile C, Del Piero F, Di Guardo G, Arispici M. Pathologic and immunohistochemical findings in naturally occurring West Nile virus infection in horses. *Vet. Pathol.* 2001; 38:414-421.

[30] Steele KE, Linn MJ, Schoepp RJ et al. Pathology of fatal West Nile virus infections in native and exotic birds during the 1999 outbreak in New York City, New York. *Vet. Pathol.* 2000; 37:208-224.

[31] Gadoth N, Weitzman S, Lehmann E. Acute anterior myelitis complicating West Nile fever. *Arch. Neurol.* 1979; 36:172-173.

[32] Platonov AE, Shipulin GA, Shipulina OY. Outbreak of West Nile infection, Volvograd region, Russia, 1999. *Emerg. Infect. Dis.* 2001; 7:128-132.

[33] Cenescu C, Ruta SM, Tardei G, et al. A high number of severe neurologic clinical forms during an epidemic of West Nile virus infection. *Rom. J. Virol.* 1997; 48:13-25.

[34] Sejvar JJ, Leis AA, Stokic DS, et al. *Acute flaccid paralysis and West Nile virus infection.* 2003; 7:788-793.

[35] Ohry A, Kaprin H, Yoeli D, Lazari A, Lerman Y. West Nile virus melitis. *Spinal Cord* 2001; 39:662-663.

[36] Joseph SA, Tsao CY. Guillain Barre syndrome. *Adolesc. Med.* 2002; 3:487-494.

[37] Leis AA, Stokic DS, Polk JL, Dostrow V, Winkelmann M. A poliomyelitis-like syndrome from West Nile virus infection. *N. Eng. J. Med.* 2002; 347:1279-1280.

[38] Glass JD, Samuels O, Rich MM. Poliomyelitis due to West Nile virus. *N. Eng. J. Med.* 2002; 347:1280-1281.

[39] Ohka S, Nomoto A. Recent insights into poliovirus pathogenesis. *Trends Microbiol.* 2001;9:501-506.

[40] Huang C, Liu C, Chang Y. Neurologic complications in children with enterovirus 71 infection. *N. Eng. J. Med.* 1999; 341:936-942.

[41] Solomon T, Kneen R, Dung N et al. Poliomyelitis-like illness due to Japanese encephalitis virus. *Lancet* 1998; 351:1094-1097.

[42] Sumarmo W, Jahja E, Gubler D, Suharyono W, Sorenson K. Clinical observations on virologically confirmed fatal dengue infections in Jakarta, Indonesia. *Bull World Health Organ* 1983; 61:693-701.

[43] Centers for Disease Control and Prevention. Acute flaccid paralysis syndrome associated with West Nile virus infection Mississippi and Louisiana, July-August 2002. *MMWR Morb. Mortal Wkly. Rep.* 2002; 51:825-828.

[44] Kelley TW, Prayson RA, Ruiz AI, Isada CM, Gordon SM. The neuropathology of West Nile virus meningoencephalitis: A report of two cases and review of the literature. *Am. J. Clin. Pathol.* 2003; 119:749-753.

[45] Ahmed S, Libman R, Wesson K, Ahmed F, Einberg K. Guillain Barre syndrome: An unusual presentation of West Nile virus infection. *Neurology* 2000; 55:144-146.

[46] Asbury AK, Cornblath DR. Assesment of current diagnostic criteria for the diagnosis of Guillain Barre syndrome. *Ann. Neurol.* 1990; 27(suppl):S21-S24.

[47] Vaispapir V, Blum A, Soboh, S, Ashkenazi H. West Nile virus meningoencephalitis with optic neuritis. *Arch. Intern. Med.* 2002; 162:606-607.

[48] Hickman SJ, Dalton CM, Miller DH, Plant GT. Management of optic neuritis. *Lancet* 2002; 360:1953-1962.

[49] Nichter CA, Pavlakis SG, Shaikh U. Rhombencephalitis caused by west Nile fever virus. *Neurology* 2002; 55:153.

[50] Armstrong RW, Pung PC. Braistem encephalitis (rhombencephalitis) due to Listeria monocytogenes: case report and review. *Clin. Infect. Dis.* 1993; 16:689-702.

[51] Hurst DL, Mehta S. Acute cerebellar swelling in varicella encephalitis. *Pediatr. Neurol.* 1988; 4:122-123.

[52] Shian WJ, Chi CS. Fatal brainstem encephalitis caused by Epstein-Barr virus. *Pediatr. Radiol.* 1994; 24:596-597.

[53] Protheroe SM, Mellor DH. Imaging in influenza A encephalitis. *Arch. Dis. Child* 1991; 66:702-705.

[54] Damasio AR. Aphasia. *N. Eng. J. Med.* 1992; 326:531-539.

[55] Kreisler A, Godefroy O, Delmaire C. The anatomy of aphasia revisited. *Neurology* 2000; 54:1117-1123.

[56] Campbell GL, Marfin AA, Lanciotti RS, Gubler DJ. West Nile virus. *Lancet Infect. Dis.* 2002; 2:519-529.

[57] Marfin AA, Gubler DJ. West Nile encephalitis: an emerging disease in the United States. *Clin. Infect. Dis.* 2001; 33:1713-1719.

[58] Perelygin AA, Scherbik SV, Zhulin IB, Stockman BM, Li Y, Brinton MA. Positional cloning of the murine flavivirus resistance gene. *Proc. Natl. Acad. Sci. USA* 2002; 9:9322-9327.

[59] Huang C, Slater B, Rudd R, et al. First isolation of West Nile virus from a patient with encephalitis in the United States. *Emerg. Infect. Dis.* 2002; 8:1367-1371.

[60] Rimland D, Koplan J, Stephen DS. Conference summary: West Nile virus southeast conference *Emerg. Infect. Dis.* 2003; 9:897-898.

[61] Southam CM, Moore AE. Clinical studies of viruses as antineoplastic agents, with particular reference to Egypt 101 virus. *Cancer* 1952; 5:1025-1034.

[62] Tardei G, Ruta S, Chitu V, Rossi C, Tsai TF, Cerenscu C. Evaluation of immunoglobulin M (IgM) and IgG enzyme immunoassays in serologic diagnosis of West Nile virus infection. *J. Clin. Microbiol.* 2000; 38:2232-2239.

[63] Hubalek Z. Comparative symptomatology of West Nile fever. *Lancet* 2001; 358:254-255.

[64] Pogodina VV, Frolova MP, Malenko GV, et al. Study on West Nile virus persistence in monkeys. *Arch. Virol.* 1983; 75:71-86.

[65] Xiao SY, Guzman H, Zhang H, et al. West Nile virus infection in the golden hamster (Mesocricetus auratus): a model for west Nile encephalitis. *Emerg. Infect. Dis.* 2001; 714-721.

[66] Solomon T, Ooi MH, Beasley DWC, Mallewa M. West Nile encephalitis. *BMJ* 2003; 326:865-869.

[67] Hayden FG. Ribavirin. In: Mandell GL, Bennet JE, Dalin ,R, eds. *Principles and practice of infectious diseases,* 5th ed. Churchill Livingston Inc, 2000; p. 477-9.

[68] Jordan I, Briese T, Fischr N, Lau JYN, Lipkin WI. Ribavirin inhibits West Nile virus replication and cytopathic effect in neural cells. *J. Infect. Dis*. 2000; 182:1214-1217.

[69] Anderson JF, Rahal JJ. Efficacy of interferon α-2b and ribavirin against West Nile virus in vitro. *Emerg. Infect. Dis*. 2002; 8:107-108.

[70] Rahal J, Anderson J, Rosenberg C, Reagan T, Thompson L. Effect of interferon alpha-2b on St Louis (SL) virus meningoencephalitis: clinical and laboratory results. *Infectious Diseases Society of America*, 40th Meeting 2002, Chicago, II:A823.

[71] Solomon T, Dung NM, Wills B et al. A double blind placebo controlled trial of interferon alpha in Japanese encephalitis. *Lancet* 2003; 361:821-826.

[72] Quick M. First treatment trial for West Nile infection begins. *Lancet Infect. Dis*. 2002; 2:589.

[73] Shimoni Z, Niven MJ, Pitlick S, Bulvik S. Treatment of West Nile virus encephalitis with intravenous immunoglobulin. *Emerg. Infect. Dis*. 2001; 7:759.

[74] Monath TP, Arroyo J, Miller C, Guirakhoo F. West Nile virus vaccine. *Curr. Drug. Targets Infect. Dsord*. 2001 1:37-50.

[75] Lustig S, Halevy M, Fuchs P, et al. Can West Nile virus outbreaks can be controlled? *Isr. Med. Assoc. J*. 2000; 2:733-737.

[76] Komar N, Panella NA, Boyce E. Exposure of domestic mammals to West Nile virus during an outbreak of human encephalitis, New York city, 1999. *Emerg. Infect. Dis*. 2001; 7:736-738.

[77] Kulasekera VL, Kramer L, Nasci RS, et al. West Nile infection in mosquitoes, birds, horses, and humans, staten island, New York, 2000 *Emerg. Infect. Dis*. 2001; 7:722-725.

[78] Petersen LR, Roering JT. West Nile virus: A reemerging global pathogen. *Emerg. Infect. Dis*. 2001; 7:611-614.

Effects of Coinfection with Borrelia Burgdorferi and Anaplasma Phagocytophilum in Vector Ticks and Vertebrate Hosts

Michael L. Levin
Centers for Disease Control and Prevention. Atlanta, Georgia, USA

Abstract

Agents of Lyme disease - *Borrelia burgdorferi* and human granulocytic anaplasmosis - *Anaplasma phagocytophilum* (formerly *Anaplasma phagocytophila*) are perpetuated in natural cycles involving the black-legged tick (*Ixodes scapularis*) and its vertebrate hosts. This predetermines the exposure of humans as well as wild and domestic animals to both pathogens and consequent concurrent infections. We studied whether a preexisting infection with either *Borrelia* or *Anaplasma* would affect acquisition and transmission of a second pathogen in ticks and their mammalian hosts. Also, we assessed the efficiency by which individual nymphs could transmit either agent alone or both agents simultaneously, to individual susceptible hosts. There was no evidence of interaction between the agents of Lyme disease and human granulocytic anaplasmosis in *I. scapularis* ticks. The presence of either agent in ticks did not affect acquisition of the other agent from an infected host. Transmission of the agents of Lyme disease and human granulocytic anaplasmosis by individual ticks was equally efficient and independent. Dually infected ticks transmitted each pathogen to susceptible hosts as efficiently as ticks infected with one pathogen only. On the other hand, a primary infection with either *B. burgdorferi* or *A. phagocytophilum* in mice inhibited acquisition and transmission of a second agent, suggesting interference between these two agents in a vertebrate host. Consequences of co-infection in ticks, wild animals and humans are discussed.

Introduction

In 1937, Evgenii Pavlovskii proposed the concept of "parasitocenosis" [parasite + Greek KOINOS - common], which describes a community of parasites and other symbionts simultaneously inhabiting a host, or even a particular host organ. Each member of a parasite community is involved in interrelationships with other members comprising the community as well as with the host organism. Direct or indirect relationships within a community may affect the fitness of the individual community members and influence the outcome of infection [87]. A concurrent infection with several pathogenic agents may result in a synergistic effect causing proliferation of individual parasite species (enhancement), or alternatively, an antagonistic interaction resulting in a reduced parasite burden (interference). Vector-borne infections provide more opportunities for pathogens to interact in parasite communities because these pathogens must survive in both vertebrate and invertebrate hosts, either or both of which may have other infections. Tick-borne pathogens, in particular, have greater opportunity for such interactions because of their ubiquity and consequent increased frequency of co-infection compared to pathogens of other arthropod vectors. Agents of Lyme disease and human granulocytic anaplasmosis are transmitted by ticks belonging to the *Ixodes persulcatus* complex. Coexistence of two pathogens within the same transmission cycle provides an exceptional opportunity to study pathogen-pathogen interactions in a naturally occurring model.

A rickettsial agent infecting granulocytes in sheep, goats and cattle was first described in Europe as a cause of tick-borne fever [45, 69] and named *Ehrlichia phagocytophila* [36]. Animals infected with the louping ill virus became sick and died of encephalitis only when simultaneously infected with the agent of tick-borne fever. Recently, it has also been recognized as a human pathogen in Europe [5, 14, 34, 68]. In Europe, this agent is transmitted by the tick *Ixodes ricinus* [69], which is also the major European vector of Lyme disease [1].

In North America, a similar agent causes granulocytic infection in humans (hence dubbed as "agent of HGE") horses (hence formerly designated *E. equi*) and dogs [6, 17, 98, 106]. A number of serologic and genetic studies have shown that the European *E. phagocytophila* and North American agent of HGE and *E. equi* belong to the same species [17, 26, 108]. Therefore, *E. phagocytophila*, *E. equi* and the HGE agent have been unified into a single species and reclassified in the family Anaplasamataceae under the genus-species name *Anaplasma phagocytophilum* [27].

In the Eastern and Midwestern United States, the black-legged tick *Ixodes scapularis* is the notorious vector of both *Borrelia burgdorferi* - etiologic agent of Lyme disease, and *A. phagocytophilum* [17, 23, 108, 110]. Both agents are maintained in a natural cycle between the tick-vector and reservoir-competent vertebrate hosts [65, 71, 85, 96, 110]. Rodents serving as important hosts for the tick species that transmits both *A. phagocytophilum* and *B. burgdorferi* have an opportunity to be coincidentally exposed to these two agents. Indeed, white-footed mice from Connecticut were shown to carry antibodies to both *A. phagocytophilum* and *B. burgdorferi* simultaneously [70], and DNA of both pathogens was detected simultaneously in a small proportion of ticks fed on white-footed mice [65], which may suggest concurrent infection with these pathogens.

Questing nymphal and adult ticks are regularly found infected with both agents simultaneously [10, 18, 20, 21, 35, 47, 50, 59, 61, 65, 67, 99-101, 104, 105, 110, 112]. Nymphal or adult ticks can acquire both-pathogens simultaneously from a single coinfected host during either larval or nymphal feeding [67]. Alternatively, adult ticks may acquire pathogens consecutively - one during larval feeding and a second during nymphal feeding. In nature, the prevalence of either pathogen in ticks increases significantly from nymphal to adult stage, and consequently, the prevalence of coinfection in questing adult ticks can be 7 to 10 X higher than in nymphs [65]. This observation suggests that consecutive acquisition of different pathogens by individual ticks may happen more frequently than simultaneous acquisition.

Simultaneous infection with these two agents has also been documented in humans and rodents [2, 7, 11, 12, 25, 51, 65, 70, 73-75, 77, 121]. Infection with both agents may theoretically result from the bite of a single coinfected tick. However, there is no experimental evidence for simultaneous transmission of *Anaplasma* and *Borrelia* by individual ticks. Moreover, the efficiency by which infected ticks can transmit either *A. phagocytophilum* or *B. burgdorferi* to susceptible hosts has not been studied in detail. A recent study found that some laboratory mice fed upon by small numbers of *I. scapularis* infected with *A. phagocytophilum* failed to acquire infection, suggesting that *A. phagocytophilum* may be transmitted less efficiently than *B. burgdorferi* [23]. However, most of the published studies on *B. burgdorferi* transmission have involved groups of infected ticks (90, 91) and it is therefore not known if all individual ticks are capable of *Borrelia* transmission. Evidence from studies of tick-borne encephalitis suggests that the efficiency of transmission by a population of infected *Ixodes persulcatus* ticks is less than 100% [55].

We measured the efficiency by which individual nymphs could transmit either agent alone or both agents simultaneously, to individual susceptible hosts. We also questioned whether previous infection with either *B. burgdorferi* or *A. phagocytophilum* in ticks would affect acquisition and/or transmission of a second pathogen. In order to determine this, we measured the efficiency with which previously infected nymphal ticks acquired a second pathogen from infected hosts and compared it to the efficiency of acquisition by uninfected ticks.

In a separate study, we investigated whether an existing infection with either *B. burgdorferi* or *A. phagocytophilum* in *P. leucopus* mice interferes with the ability of mice to acquire and transmit a second infection to ticks, consequently reducing their role in the natural maintenance of the second agent.

1. Coinfection with Borrelia Burgdorferi and Anaplasma Phagocytophilum in Vector Ticks

The white-footed mouse (*Peromyscus leucopus*) is known to be a major reservoir for *B. burgdorferi*. It also has been shown susceptible to infection with *A. phagocytophilum* [24, 65, 67, 81, 110]. Therefore, we used white-footed mice as hosts in our experiments. Two-month-old mice were derived from a specific pathogen-free *P. leucopus* colony maintained in the Vector Ecology Laboratory at Yale School of Medicine (New Haven, CT).

The maintenance and care of experimental animals complied with the National Institutes of Health guidelines for the humane use of laboratory animals. Mice were not exposed to ticks or pathogens prior to the experiments.

Infected *I. scapularis* nymphs were produced by allowing larval ticks to feed upon white-footed mice previously infected with either *B. burgdorferi* or *A. phagocytophilum*. Both agents originated from nymphal ticks collected in Westchester Co. (NY) and were maintained separately in a laboratory tick - mouse cycle. Identity of the agents had been previously confirmed by indirect immunofluorescence assay and by DNA sequencing of amplified PCR products [23, 63].

Infection with *B. burgdorferi* and *A. phagocytophilum* in ticks and mice was determined using PCR. Individual nymphal or adult ticks, or pools of engorged larvae were placed in sterile 1.5 cc plastic vials, deep-frozen in liquid nitrogen, ground with sterile plastic pestle, and re-suspended in 100 µl of TB buffer. DNA was extracted from ticks using IsoQuick Nucleic Acid Extraction Kit (ORCA Research Inc., Bothell, WA) to maximize sensitivity [102]. Briefly, guanidine thiocyanate, a proprietary extraction matrix and sodium dodecyl sulfate solution were added to a suspension, and the mixture was incubated at 65°C for 10 min. After separation of phases by centrifugation, DNA was precipitated with sodium acetate and isopropanol and washed with 70% ethanol. The final DNA pellet was resuspended in 50 µl of RNase -free water, and 2.5 µl aliquot was used for each PCR test Primers EHR521 (5-TGT AGO CGG TTC GOT AAG TTA AAG-3') and EHR747 (5--GCA CTC ATC GTT TAC AGC GTG-3') were used to amplify a 247-bp fragment of 16S rDNA from *A. phagocytophilum* [85]. Primers FLA297 (51-CGG CAC ATA TTC AGA TGC AGA CAG-3') and FLA652 (5'-CCT GTT GAA CAC CCT CTT GAA CC-3') based on the published nucleotide sequence [39] were used to amplify a 378-bp fragment of the flagellin gene of *B. burgdorferi*. The amplification products were visualized in 2% agarose gels.

1.A. Acquisition Experiment

Ten mice were each infected with *Borrelia* by feeding with 10 *I. scapularis* nymphs from a *B. burgdorferi* -infected cohort. Another 10 mice were each similarly infected with *Anaplasma* by feeding 10 nymphs from an *A. phagocytophilum* - infected cohort. The infection in nymphal cohorts prior to the investigation was assessed by testing representative samples of 25 ticks. Prevalence of infection in *Borrelia* -infected cohort was 44.0±10.1%, and prevalence of infection in *Anaplasma* -infected cohort was 40.0±10.0%.

Two weeks later, 25 nymphs from the *B. burgdorferi* -infected cohort were placed on each of 5 mice previously infected with *Anaplasma*. The other 5 *Anaplasma* -infected mice were each fed upon by 25 uninfected nymphs. Similarly, *B. burgdorferi* -infected mice were infested with 25 nymphs from the *Anaplasma* - infected cohort, and 25 uninfected nymphs fed upon the other 5 *B. burgdorferi* -infected mice. Engorged nymphs were collected and kept at 22°C and 98% relative humidity until molt. Freshly molted adult ticks were individually tested for infection by PCR.

An average of 19 [12 to 23] nymphal ticks fed to repletion on each of the 20 infected mice, and were tested for both agents as adults. When nymphs from the *Borrelia*-infected cohort fed upon five mice infected with *Anaplasma*, 39 of the resulting adult ticks tested PCR-positive, and 47 tested PCR-negative for *B. burgdorferi* (Table 1). Prevalence of *Borrelia* infection in adult ticks (45.3±10.6%) did not differ from that in the same cohort of nymphs tested prior to feeding (44.0±19.9%). When nymphs from the *Anaplasma*-infected cohort fed upon five mice infected with *B. burgdorferi*, 48 of the resulting adult ticks were tested PCR-positive, and 47 were tested PCR-negative for *Anaplasma* (Table 2). Again, the difference in *Anaplasma* infection between nymphal ticks prior to feeding (40.0±19.6%) and the resulting adult ticks (50.5±10.1%) was not statistically significant.

Table 1. Acquisition of *A. phagocytophilum* by ticks infected with *B. burgdorferi* and by uninfected ticks

Mouse #	Ticks infected with *B. burgdorferi*			Uninfected ticks			$P_{\chi 2}$*
	Tested	Acquired *A. phagocytophilum* (%)		Tested	Acquired *A. phagocytophilum* (%)		
Pl-387	11	6	(54.5)	9	4	(44.4)	0.65
Pl-388	6	3	(50.0)	6	3	(50.0)	1.00
Pl-389	8	5	(62.5)	12	8	(75.0)	0.85
Pl-390	7	4	(57.1)	7	3	(42.9)	0.59
Pl-391	7	2	(28.6)	13	5	(38.5)	0.67
Total	39	20	(51.3±15.9) **	47	24	(51.1±14.4) **	0.98

* $P_{\chi 2}$ – the probability for a chi-square test distribution.
** ± 95% Confidence interval.

Nymphs may be able to acquire pathogens not only from an infectious host, but also from infected ticks during cofeeding [41, 82, 83, 86]. However, transmission by cofeeding did not increase the prevalence of either *B. burgdorferi* or *Anaplasma* in our experiment. Therefore, we assume that the same ticks, which tested positive for *B. burgdorferi* after feeding upon mice (Table 1), were indeed infected with *B. burgdorferi* prior to feeding. The same assumption applies to the adult ticks that tested positive for *Anaplasma* (Table 2).

Table 2. Acquisition of B. burgdorferi by ticks infected with A. phagocytophilum and by uninfected ticks

Mouse #	Ticks infected with *A. phagocytophilum*			Uninfected ticks			$P_{\chi 2}$*
	Tested	Acquired *B. burgdorferi* (%)		Tested	Acquired *B. burgdorferi* (%)		
Pl-397	6	4	(66.7)	14	10	(71.4)	0.83
Pl-398	8	7	(87.5)	9	7	(77.8)	0.60
Pl-399	11	11	(100.0)	7	7	(100.0)	1.00
Pl-400	12	11	(91.7)	8	8	(100.0)	0.40
Pl-401	11	8	(72.7)	9	8	(88.9)	0.39
Total	48	41	(85.4±10.0)**	47	40	(85.1±10.3)**	0.97

* $P_{\chi 2}$ – the probability for a chi-square test distribution.
** ± 95% Confidence interval.

A total of 44 ticks from the *Borrelia*-infected cohort acquired *Anaplasma* during feeding upon 5 infected mice (Table 1). The efficiency of *Anaplasma* acquisition by nymphal ticks from the same cohort varied between individual mice, but did not differ between ticks that were or were not previously infected with *B. burgdorferi* ($P_{\chi 2} = 0.98$). On the average, approximately 50% of nymphs acquired *Anaplasma* from infected mice regardless of their prior infection status with *B. burgdorferi* (Table 1).

When a cohort of exclusively uninfected nymphs fed upon the second group of 5 mice infected with *Anaplasma*, a total of 43 of 96 resulting adult ticks (44.8±10.0%) acquired the infection. Individual mice transmitted *Anaplasma* to 30.0 to 60.1% of feeding ticks. The difference in acquisition of *Anaplasma* by a cohort of *B. burgdorferi*-infected nymphs and a cohort of uninfected nymphs was not statistically significant ($P_{ANOVA} = 0.22$).

A total of 81 ticks from the *Anaplasma*-infected cohort acquired *Borrelia* during feeding upon 5 infected mice (Table 2). The efficiency of *B. burgdorferi* acquisition by nymphal ticks from the same cohort varied between individual mice, but did not differ between ticks that were or were not previously infected with *Anaplasma* ($P_{\chi 2} = 0.97$). An average of 85.3±7.2% of nymphs acquired *B. burgdorferi* from infected mice regardless of their prior infection status with *Anaplasma* (Table 2).

When a cohort of exclusively uninfected nymphs fed upon five additional mice infected with *B. burgdorferi*, a total of 88 of 105 resulting adult ticks (83.8±7.1%) acquired the infection. Individual mice transmitted *B. burgdorferi* to 73.9 to 90.5% of feeding ticks. The difference in acquisition of *B. burgdorferi* by a cohort of *Anaplasma*-infected nymphs and a cohort of uninfected nymphs was not statistically significant ($P_{ANOVA} = 0.37$).

Thus, previous infection with *B. burgdorferi* or *A. phagocytophilum* in nymphal *I. scapularis* did not affect the ability of ticks to acquire a second pathogen from infected hosts.

1.B. Transmission Experiment

Single-infected and coinfected nymphs were produced by allowing larval ticks to feed upon white-footed mice singly or simultaneously infected with *B. burgdorferi* and *A. phagocytophilum* in the course of the previous experiment. These nymphs were placed individually on single naive mice and allowed to feed to repletion. The resulting engorged nymphs were collected and individually tested for infection by PCR.

Two weeks after the feeding by infected nymphs, the mice were infested with uninfected larval ticks for xenodiagnosis. The infection status of individual mice was assessed using 20 engorged xenodiagnostic larvae per mouse (4 pools of 5 ticks). Tick pools were tested for both pathogens by PCR. Our previous study had shown that feeding density influences acquisition of *B. burgdorferi* in larval *I. scapularis* [64]. Therefore, mice were infested with a large number of larvae (approximately 200) in order to maximize the sensitivity of xenodiagnosis. Xenodiagnostic larvae were derived from a colony of *I. scapularis* maintained in the Vector Ecology Laboratory at Yale School of Medicine by feeding on uninfected mice and rabbits for several generations. Representative samples of ticks from the colony are regularly tested to ensure that the colony is free of both tick-borne pathogens. Xenodiagnosis was performed only on mice from which individual replete nymphs were collected and tested positive for either pathogen.

A total of 98 mice were successfully fed upon by individual nymphal ticks. Of those 98 nymphs, 89 were infected with either *B. burgdorferi*, or *A. phagocytophilum* or both as detected by PCR performed on engorged ticks (Table 3). Xenodiagnostic results showed that 31 of 38 (81.6%) ticks infected with *B. burgdorferi* only transmitted the spirochete to mice, compared to 70% (21 of 30) transmission success when ticks were simultaneously infected with both *Borrelia* and *Anaplasma* (Table 3). This difference between the 2 groups of ticks was not statistically significant ($P_{\chi 2} = 0.27$). When ticks were infected with *Anaplasma* only, 18 of 21 (85.7%) transmitted it to susceptible mice, as determined by xenodiagnosis (Table 3). Of 30 dually infected ticks, 25 (83.3%) transmitted *Anaplasma*. Thus, there also was no difference in the efficiency of transmission of *A. phagocytophilum* between ticks infected with one or both pathogens ($P_{\chi 2} = 0.82$). Efficiency of transmission did not differ significantly between *B. burgdorferi* and *Anaplasma*, neither in ticks infected with one pathogen ($P_{\chi 2} = 0.69$), nor in dually infected ticks ($P_{\chi 2} = 0.22$).

Table 3. Transmission of *B. burgdorferi* and *A. phagocytophilum* to mice by individual *Ixodes scapularis* nymphs

Infection in nymphs	Ticks fed	Transmitted *B. burgdorferi* No. (%)*	Transmitted *A. phagocytophilum* No. (%)*
B. burgdorferi only	38	31 (81.6 ± 12.5)	--
B. burgdorferi & *A. phagocytophilum*	30	21 (70.0 ± 16.7)	25 (83.3 ± 13.6)
A. phagocytophilum only	21	--	18 (85.7 ± 15.3)

* ± 95% Confidence interval.

In another study on pathogen transmission by individual *I. scapularis*, 6 of 7 nymphs that fed to repletion transmitted *B. burgdorferi* to hamsters [89]. However, ticks themselves were not examined, and it was not known whether non-transmitting ticks were infected. Our data show that only 70 to 81% of infected *I. scapularis* nymphs transmit *B. burgdorferi* to susceptible hosts even when fed to repletion. The efficiency of transmission of *A. phagocytophilum* by infected ticks is 83 to 86%, and is not significantly different from *B. burgdorferi*. Differential infectivity of ticks is likely to be related to the variability of pathogen concentration between infected ticks [15, 54, 56, 57, 66]. This has been shown to occur in ticks transmitting spring-summer tick-borne encephalitis virus [52, 53].

Table 4. Prevalence of *B. burgdorferi* and *A. phagocytophilum* in mice and xenodiagnostic I. scapularis ticks fed as uninfected larvae upon mice during 5 consecutive infestations

Mice exposed to:	Ticks tested*	*B. burgdorferi*		*A. phagocytophilum*	
		Infected mice/N	Prevalence in ticks (%±SE)	Infected mice/N	Prevalence in ticks (%±SE)
Borrelia only (Group I)	342	5/5	81.6±1.7	0/5	0.0
Anaplasma + *Borrelia* (Group II)	457	5/5	59.6±2.1	5/5	27.7±2.0
Anaplasma only (Group III)	533	0/5	0.0	5/5	21.2±1.8
Borrelia + *Anaplasma* (Group IV)	459	5/5	77.8±1.8	2/5	10.5±1.4

* All ticks in each group were individually tested for both agents.

Of 30 ticks infected with *Borrelia* and *Anaplasma* simultaneously, 20 successfully transmitted both pathogens, while 4 failed to transmit either (Figure 1). Our results suggest that transmission of the agents of Lyme disease and HGE by individual ticks is equally efficient and independent. Simultaneous infection with the agents of Lyme disease and HGE has been observed both in human patients and in wild animals [2, 7, 11, 12, 25, 51, 65, 70, 73-75, 77, 121]. Mixed infections in hosts may originate either from the bite of a single tick infected with two pathogens, or from multiple bites of singly infected ticks. Simultaneous transmission of *B. burgdorferi* and *Babesia microti* by individual *I. scapularis* nymphs has been previously reported [89]. The present study provides conclusive evidence that dually infected ticks are capable of simultaneous transmission of *B. burgdorferi* and *A. phagocytophila*, and that concurrent infection in humans can indeed be caused by the bite of a single coinfected tick.

However, the prevalence of dual infection in nymphs in nature is 4-5 times lower than that of a single infection with either *Borrelia* or *Anaplasma* [65, 101]. We reasoned that under natural conditions vertebrate animals were more likely to encounter these agents sequentially from multiple tick bites than simultaneously from a single co-infected tick.

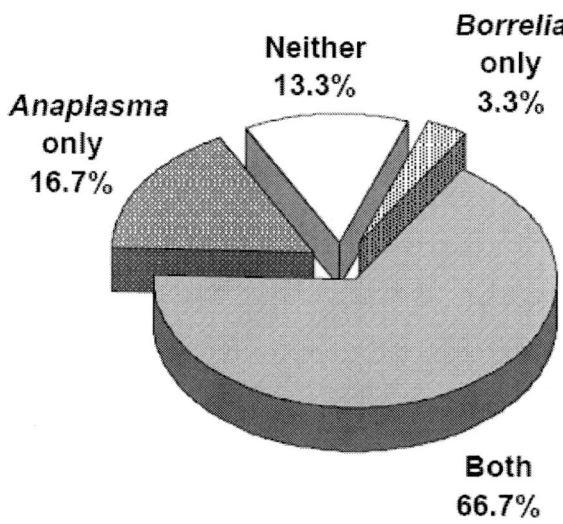

Figure 1. Transmission of *Borrelia burgdorferi* and *Anaplasma phagocytophilum* by individual *Ixodes scapularis* nymphs simultaneously infected with both pathogens.

2. Interaction Between *Borrelia Burgdorferi* and *Anaplasma Phagocytophilum* in Sequential Infection in a Reservoir Host

In order to study reciprocal effects of the two pathogens in sequential infections, we first infected laboratory bred 4 wk.-old *P. leucopus* mice with either *B. burgdorferi* or *A. phagocytophilum*, and subsequently challenged them with the other agent. The susceptibility of pre-infected mice to the second infection and their subsequent infectivity for xenodiagnostic ticks were compared to the same parameters in uninfected control mice. Strains of *B. burgdorferi* and *A. phagocytophilum* used in our experiments originated from *I. scapularis* nymphs collected at an endemic site in Westchester Co. (NY) [101]. The agents were maintained in a tick-mouse cycle, where infected *I. scapularis* nymphs were produced by allowing uninfected larval ticks to feed upon mice previously exposed to infected nymphs. Uninfected xenodiagnostic larvae were derived from a separate *I. scapularis* colony maintained for several generations by feeding on uninfected mice (immatures) and rabbits (adults).

Primary Infection

Twenty tick-naive and specific pathogen-free *P. leucopus* mice were randomly assigned to 4 groups (Figure 2). To establish primary infections, two groups of mice were exposed to *I. scapularis* nymphs infected with either *A. phagocytophilum* or *B. burgdorferi* one week prior to the challenge. Mice in one group were each infested with 10 nymphs with 45% prevalence of *Anaplasma* (Group II, Figure 2), and mice in another group were similarly infested with 10 nymphs with 70% prevalence of *Borrelia* (Group IV, Figure 2). Mice in the two control groups remained uninfected until the day of challenge, which was defined as Day 0 (Groups I and III, Figure 2).

Challenge and Xenodiagnosis

On Day 0, five uninfected mice (Group I) and 5 *Anaplasma*-pre-infected mice (Group II) were each challenged with 10 *Borrelia*-infected nymphal ticks. Simultaneously, another 5 uninfected mice (Group III) and 5 *Borrelia*-pre-infected mice (Group IV) were challenged with 10 nymphs infected with *Anaplasma* (Figure 2). Approximately 100 uninfected *I. scapularis* larvae were placed on each mouse on Days 7, 14, 21, 28, and 58 for xenodiagnosis. For the duration of the experiment, mice were kept in individual wire-mesh cages over water. Engorged larvae were collected daily as they detached after repletion and kept in an environmental chamber at 22°C and 95% relative humidity until they molted to the nymphal stage. A sample of 15 to 25 nymphs derived from xenodiagnostic larvae per mouse per infestation were individually tested by PCR for both pathogens within 3-5 wk after molting.

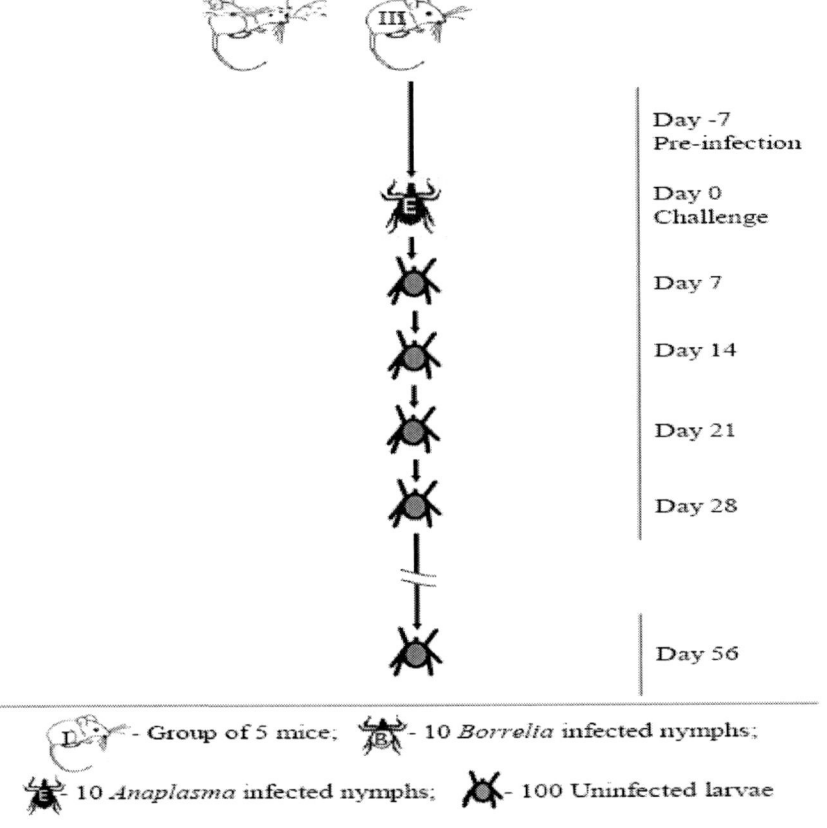

Figure 2. Scheme of the experiment on sequential infection with *B. burgdorferi* and *A. phagocytophilum* in white-footed mice.

Xenodiagnosis is the most sensitive method for detection of *Borrelia* infection in mice [72, 79, 103], but the sensitivity of xenodiagnosis for identifying infection with *Anaplasma* was yet to be determined. Therefore, in addition to xenodiagnosis, PCR of whole blood and indirect immunofluorescence assay (IFA) of sera were used to confirm *Anaplasma* infection in mice. Blood samples for PCR were collected from the retro-orbital sinus of all mice on the day of each infestation. Sera samples for IFA were collected at euthanasia, 63 days after the challenge.

PCR

Mouse blood and ticks were tested for the presence of *B. burgdorferi* and *A. phagocytophilum* DNA as described above (this chapter).

IFA

Serum samples were screened for the presence of specific IgG antibodies against *A. phagocytophilum* using the previously described method [80]. The antigen was derived from human promyelocyte cell culture (HL-60) infected with the agent of HGE obtained from Westchester County, NY. Sera were screened at a dilution of 1:40 in 1X phosphate-buffered saline (pH 7.4) [65].

F and T statistical analyses were applied to *arcsine*-transformed ratios to compare prevalence of infection in xenodiagnostic ticks.

Borrelia Only (Group I)

All uninfected mice acquired *B. burgdorferi* when challenged by *Borrelia*-infected nymphs and became infectious for feeding larvae. Prevalence of infection in xenodiagnostic ticks was higher than 80% in ticks fed during the 4 weeks following the challenge. Prevalence peaked at 91.8% (± S.E. 3.4%) at 3 weeks after the challenge, and decreased to 51.7±16.1% by 8 weeks after the challenge (Figure 3). Over the course of 5 consecutive infestations, mice infected with only *Borrelia* transmitted the spirochete to 81.6±1.7% of xenodiagnostic ticks (Table 2, Group I). None of the ticks in this group tested positive for *Anaplasma*. Blood PCR and IFA results for *Anaplasma* in mice were also negative.

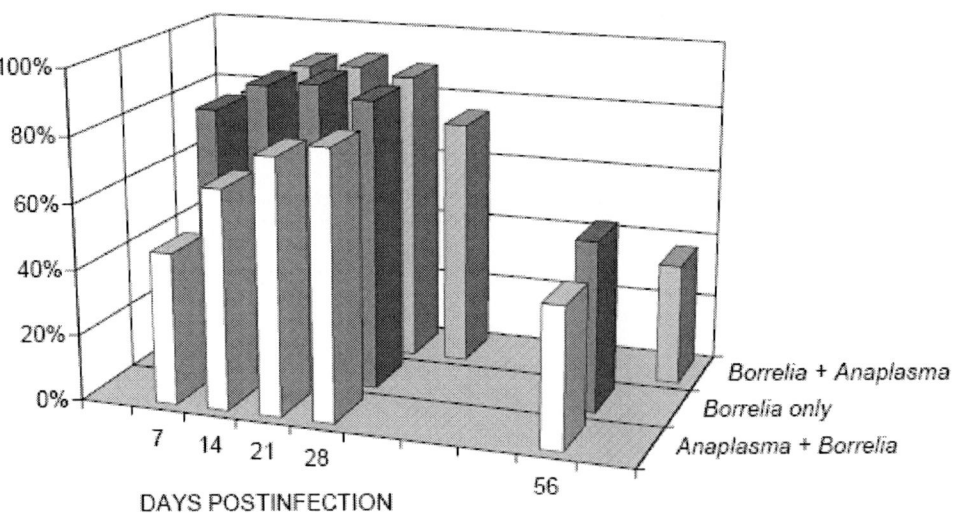

Figure 3. Acquisition of *B. burgdorferi* by *I. scapularis* larvae from mice infected with 1 or 2 pathogens in different sequence.

Anaplasma + *Borrelia* (Group II)

Five mice that were exposed to *Anaplasma*-infected nymphs 1 wk before the challenge all acquired *A. phagocytophilum*. They were blood-PCR positive at 1 and 2 weeks postinfection (Days 0 and 7 respectively), and all developed IgG antibodies to *Anaplasma* by the end of the experiment. All 5 mice also transmitted *Anaplasma* to xenodiagnostic ticks. Overall, 27.7±4.1% of ticks became infected with *Anaplasma* after feeding on these mice over 5 xenodiagnostic infestations (Table 2, Group II).

When challenged by *Borrelia*-infected nymphs, all *Anaplasma*-infected mice acquired *Borrelia* as well and transmitted *Borrelia* to xenodiagnostic ticks. The average prevalence of *Borrelia* in ticks fed upon these dually infected mice gradually increased from 46.2±18.6% on Day 7 to 81.5±5.8% on Day 28, and decreased to 42.5±13.9% by Day 56 (Figure 3). Overall, only 62.5±3.2% of ticks became infected with *Borrelia* after feeding on these mice over 5 xenodiagnostic infestations (Table 2, Group II), and 19.6±4.6% of ticks had mixed infections.

Anaplasma Only (Group III)

All uninfected mice in Group III also successfully acquired *Anaplasma* when challenged by *Anaplasma*-infected nymphs. All five mice were blood-PCR positive at 1 and 2 weeks postinfection (Days 0 and 7 respectively), and developed IgG antibodies to *Anaplasma* by the end of the experiment. Over the course of 5 consecutive infestations, infected mice transmitted *Anaplasma* to 21.2±1.8% of xenodiagnostic ticks (Table 2, Group III). Prevalence of *Anaplasma* in xenodiagnostic ticks was highest (48.7±5.8) in ticks fed on Day 7. It gradually decreased to 9.9±2.7% by Day 28, and to 2.0±1.3% by 8 weeks after the challenge (Figure 4). None of the ticks in this group tested positive for *Borrelia*.

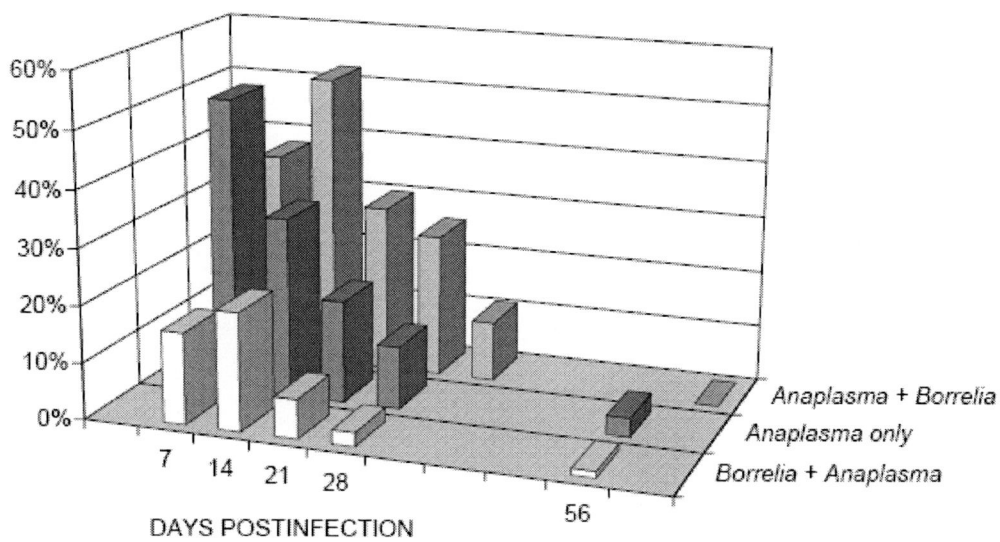

Figure 4. Acquisition of *A. phagocytophilum* by *I. scapularis* larvae from mice infected with 1 or 2 pathogens in different sequence.

Borrelia + Anaplasma (Group IV)

Mice that were exposed to *Borrelia*-infected nymphs 1 wk before *Anaplasma* challenge all became infected with *Borrelia* as confirmed by xenodiagnosis. Overall, 77.8±1.8% of ticks acquired *Borrelia* after feeding on these mice over 5 xenodiagnostic infestations (Table 2, Group IV).

Following the challenge with *Anaplasma*-infected nymphs, all five *Borrelia*- infected mice developed IgG antibodies to *Anaplasma*. However, *Anaplasma* DNA was detected in xenodiagnostic ticks and in blood samples from only two of five mice. During 5 consecutive xenodiagnostic infestations, mice in this group transmitted *Anaplasma* to only 10.5±1.4% of feeding ticks (Table 2, Group IV), and 8.5±1.3% of ticks had dual infection. Prevalence of *Anaplasma* infection was highest (20.8±4.1%) in ticks fed on Day 14 (Figure 4). *Anaplasma* DNA was detected in the blood of one mouse on Day 7 after challenge and in the blood of a second mouse on Day 14. Although all five *Borrelia*- infected mice responded to *Anaplasma* challenge by producing antibodies, only two transmitted the secondary infection to feeding ticks. The other 3 mice, apparently, had either a brief *Anaplasma* infection or an infection insufficient for detection by either blood-PCR or xenodiagnosis. Xenodiagnostic ticks feeding on these mice acquired *Borrelia* only.

In the *Borrelia* challenge experiment, all mice, both naive and *Anaplasma*-infected, acquired *Borrelia* and infected xenodiagnostic ticks with *Borrelia*. However, the efficiency of transmission differed. Mice that were infected with only *Borrelia* transmitted to significantly higher proportion of ticks than did mice infected with both agents (P=<0.01). Similarly, mice that were infected with only *Anaplasma* transmitted to significantly higher proportion of ticks than did those infected both agents (P=<0.01).

We also analyzed how mixed infection affected the acquisition of *Borrelia* by feeding larval ticks over time. Prevalence of *Borrelia* infection in xenodiagnostic ticks fed upon mice infected with only *Borrelia* (Group I) fluctuated between 80% and 92% during the first 28 days postinfection, and declined to about 52% after 2 months (Figure 3). The mean prevalence of *Borrelia* infection in xenodiagnostic ticks fed upon mice previously infected with *Anaplasma* (Group II) was lower compared to ticks fed upon mice infected with only *Borrelia* throughout the experiment. This difference was statistically significant for xenodiagnostic ticks fed at 7, 14, and 21 days after the challenge with *Borrelia*-infected nymphs ($P=<0.01$, $P=0.05$, and $P=<0.05$ respectively), and became marginal ($P > 0.1$) in ticks fed at 28 and 56 days post-challenge. Thus, development of *Borrelia* infectivity for xenodiagnostic ticks in mice previously infected with *Anaplasma* was significantly delayed compared to *Borrelia* infection alone. However, the acquisition of *Borrelia* by ticks was not altered when *Anaplasma* was introduced into mice having an existing *Borrelia* infection (Figure 3).

Similarly, we analyzed how mixed infection from this experiment affected acquisition of *Anaplasma* by feeding larval ticks over time. Mean prevalence of *Anaplasma* in xenodiagnostic ticks fed upon mice infected with only *Anaplasma* was highest at 7 days postinfection, and gradually declined thereafter (Figure 4). Mice that had an existing infection with *Borrelia* and were later exposed to *Anaplasma* exhibited a somewhat similar pattern, though their ability to infect ticks was lower. The difference in the prevalence of *Anaplasma* was statistically significant for xenodiagnostic ticks fed at 7, 14, 21 and 28 days after the challenge with *Anaplasma*-infected nymphs ($P=<0.01$, $P=0.05$, $P=<0.05$, and $P=<0.05$ respectively). Conversely, in *Anaplasma*-infected mice, challenge with *Borrelia* seemed to increase transmission of *Anaplasma* to feeding ticks, though the difference was not statistically significant. Average prevalence of *Anaplasma* in ticks fed at comparable intervals on mice challenged with *Borrelia* was 12-15% higher compared to singly infected mice starting 7 days after the challenge.

Discussion

Thus, individual *I. scapularis* nymphs derived from larvae infected with either *B. burgdorferi* or *A. phagocytophilum* remained equally susceptible to a second pathogen. *I. scapularis* nymphs coinfected with *B. burgdorferi* and *A. phagocytophilum* are capable of transmitting both agents to white-footed mice; i.e. mice are as susceptible to dual infections transmitted simultaneously as to individual infections with *B. burgdorferi* or *A. phagocytophilum*. Apparently, both acquisition and transmission of infection occurs independently in ticks with multiple infections.

On the other hand, the prevalence of dual infection in nymphs in nature is 4-10 times lower than that of a single infection with either *Borrelia* or *Anaplasma* [10, 61, 65, 85, 100, 101, 110, 112], and only a small proportion of ticks pose a risk of true coinfection in most regions. We reasoned that under natural conditions vertebrate animals were more likely encounter these agents sequentially from multiple tick bites than simultaneously from a single co-infected tick bite. Accordingly, we also studied outcomes of reciprocal sequential infections of *Borrelia* and *Anaplasma* in mice.

In our experiments, all white-footed mice that were initially infected with *Anaplasma* remained susceptible to a secondary infection with *Borrelia* when exposed to infected ticks. However, these dually infected mice infected fewer xenodiagnostic ticks with *Borrelia* compared to mice infected with *Borrelia* only. There was no evidence of increased severity of *Borrelia* infection resulting from co-infection with *Anaplasma*. In fact, our results suggest that the development of *Borrelia* was suppressed in mice previously infected with *Anaplasma*. Also, introduction of *Anaplasma* into mice already infected with *Borrelia* had little or no effect on their ability to transmit *Borrelia* to larval ticks.

All mice infected with only *Anaplasma* transmitted infection to xenodiagnostic ticks. Conversely, *Anaplasma* was detected in xenodiagnostic ticks in only 2 of 5 mice previously infected with *Borrelia* ($P=<0.01$). Therefore, mice that were initially infected with *Borrelia* were less susceptible to a secondary infection with *Anaplasma*. It appears that a primary infection with *Borrelia* prevented or suppressed development of a secondary *Anaplasma* infection iYi3 of 5 mice. This resulted in an overall lower prevalence of *Anaplasma* infection in ticks fed upon *Borrelia* pre-infected mice compared to mice either infected with *Anaplasma* only (Group III) or infected with *Anaplasma* first (Group II).

Thus, a primary infection with either *Borrelia* or *Anaplasma* in white-footed mice inhibited the maintenance of a second agent in a tick-mouse cycle. These findings are markedly different from results of the study on coinfection in ticks (this chapter), where *I. scapularis* nymphs coinfected with *B. burgdorferi* and *A. phagocytophilum* successfully transmitted both agents to *P. leucopus*. They also contrast sharply with those in a study on *Borrelia* and *Anaplasma* coinfection in laboratory mice [111]. These authors reported that simultaneous needle-inoculation of C3H/HeN and C3H/HeN-*scid* mice with cultures of *B. burgdorferi* and *A. phagocytophilum* resulted in increased bacterial burden, Lyme arthritis, and pathogen transmission to the vector. The apparent differences between studies with simultaneous and sequential infection indicate that outcomes of coinfection in mammalian hosts may vary depending on whether *B. burgdorferi* and *A. phagocytophilum* are introduced simultaneously or sequentially, notwithstanding the fact that Thomas with coauthors used in their study animals with altered or severely compromised immune responses, an artificial rout of infection, and an arbitrary infectious dose.

Similar "contradicting" results were obtained in studies on coinfections with *Francisella tularensis* and *Listeria monocytogenes*, *F. tularensis* and *Salmonella typhimurium*, *F. tularensis* and *Staphylococcus aureus* in hares, guinea pigs, voles, and laboratory rats and mice where reactions to mixed infections depended strongly on the interval between inoculations [29-31, 92]. For example, *F. tularensis* was detected in blood of 100% of common voles (*Microtus arvalis*) when inoculated simultaneously with *S. typhimurium*, but in only 23-37% of those infected with *Francisella* at 4-7 days after *Salmonella* [33]. *Listeria* was successfully reisolated from 75% of laboratory mice if they were inoculated with the two pathogens simultaneously, but only from 29% when *F. tularensis* was introduced at a four day interval. In experiments with longer intervals between inoculation of *Listeria* and *F. tularensis*, reisolation of the first pathogen was successful from only 1.4% of mice. Likewise, 100% of guinea pigs died of tularemia if *F. tularensis* was inoculated either simultaneously with *Listeria* or after a 10-13 day interval, but two out of 10 guinea pigs survived when *F. tularensis* was inoculated on day 4 following *Listeria* inoculation. In addition, dually infected animals survived longer than those infected with the tularemia agent only [31, 32]. It is evident that alteration of dosages or the order and intervals between introductions of the same pathogens into the same vertebrate host species can bring forth different and sometimes opposite effects, whereupon coinfection may either stimulate or suppress the host defense mechanisms. Moreover, similar combinations of the same pathogens in different species of hosts may also produce totally different outcomes.

Disparate results of mixed infection between ticks and mice suggest that interactions between *Borrelia* and *Anaplasma* in vertebrates are indirect and likely mediated by host immune responses. Infection with *A. phagocytophilum* had been associated with false-positive serological tests (ELISA) for *B. burgdorferi* infection in humans [119, 120]. Patients with documented HGE and no symptoms of early Lyme disease exhibited serologic reactivity to *B. burgdorferi*. Also, BALB/c mice infected with *A. phagocytophilum* produced antibodies cross-reactive with *B. burgdorferi* proteins migrating in the region of OspC, p39, and p83 [49, 78]. These observations suggest that cross-reactivity between two pathogens may play a role in the suppression of a secondary *B. burgdorferi* infection in white-footed mice previously infected with *A. phagocytophilum*. Another possibility is that a somewhat lesser degree of susceptibility to a sequentially introduced pathogen may result from nonspecific stimulation of the host immune system by the first infection.

A. phagocytophilum infections are often reported to result in immunosuppression. Granulocytic ehrlichiosis was first described as a cofactor for a disease in the livestock infected with the louping ill virus [45] [109]. Since then, it has been shown that *A. phagocytophilum* infection in sheep, goats, and humans result in multifactorial defects in the host immune defense including neutropenia, thrombocytopenia, decreased neutrophil adherence, emigration, and phagocytosis, decreased production of antibodies, and decreased lymphocyte mitogenesis [8, 37, 38, 43, 60, 97, 113, 115-118]. These pathogen-mediated changes in host defense and immune suppression can allow secondary and opportunistic infections. In fact, major clinical manifestations of granulocytic ehrlichiosis in sheep do not appear to be due to the infection with *A. phagocytophilum* itself, but to secondary infections [13]. Several viral, bacterial, and fungal agents normally nonpathogenic or only mildly pathogenic to animals cause severe and fatal illnesses due to a concurrent infection with *A. phagocytophilum* [9, 13, 42, 44, 76, 84]. It could have been expected therefore, that the competence of natural hosts, including *P. leucopus,* as reservoirs for *Borrelia* also may be affected by coinfection with *A. phagocytophilum*, and an increasing number of spirochetes in tissues and blood could, in turn, augment the infection prevalence in feeding ticks [88].

There are, however, observations that do not support the role of *A. phagocytophilum* as a universal facilitator of secondary infections. Simultaneous infection with *A. phagocytophilum* and *Babesia divergens* in cattle does not induce proliferation of either pathogen [16]. Moreover, it has been reported that *A. phagocytophilum* infection tends to delay or even completely suppress the establishment of *B. divergens* if inoculated simultaneously or one week previously [93, 94]. Larsen [60] found that immunization with *Actinomyces pyogenes* resulted in significantly higher antibody titers in *A. phagocytophilum*-infected sheep compared to the control group, as measured by an enzyme-linked immunosorbent assay (ELISA). Our results indicate that *B. burgdorferi* may be added to a list of pathogenic microorganisms that do not necessarily benefit from coinfection with *A. phagocytophilum*.

Existence of both *B. burgdorferi* and *A. phagocytophilum* in nature is solely dependent on the ability of available vertebrate hosts to acquire infection and to transmit it to a new cohort of ticks. *I. scapularis* and *I. ricinus* ticks are ambushing parasites and do not actively move far from the place of hatching or molting while questing for a new host. A prospective host must be moving actively to collect these ticks, and the most mobile individuals within a species accumulate larger numbers of ticks [3]. Thus, survival of a tick-borne pathogen largely relies on the mobility of a reservoir host. There are no studies published on changes in behavior or physiology of *P. leucopus* due to mixed infection versus single infection with either pathogen. It is easy to envision, however, that if the activity of reservoir hosts becomes suppressed because of concurrent infections with several pathogens, they would harbor fewer ticks, and infectious agents would have a decreased probability of being transmitted to a new cohort of vectors. One way a tick-borne pathogen can "ensure" its successful transmission would be suppression or elimination of subsequent infections that may result in decreased contact between host and vector ticks. Therefore, there may be a selective advantage for agents transmitted by the same tick-vector to exhibit antagonistic relationships in natural reservoir hosts. Mechanisms by which an existing infection may suppress transmission of competing vector-borne agents are yet to be studied.

Like in wild and domestic animals, human coinfection may result from a bite of a single tick carrying multiple pathogens or from multiple bites at different points in time. True coinfection should be distinguished from sequential infection as clinical manifestations of a simultaneous infection may not be seen in patients with sequential infection. Simultaneity of introduction of two or more pathogens is usually difficult to establish in an uncontrolled situation, unless coinfection is documented both in a patient and in the infecting tick. Case reports are useful to confirm the occurrence of mixed infections and describe the clinical and laboratory characteristics. Although many of these reports tend to focus on severe or unusual cases [2, 11, 25, 73, 75, 77], existing data do not seem to support the increased severity of Lyme disease with HGE beyond the typical clinical manifestations [4, 58, 75]. Coinfected patients tend to have more symptoms, but their clinical pictures lack distinctive characteristics [107]. The frequency of clinically confirmed cases of simultaneous infection with *B. burgdorferi* and *A. phagocytophilum* is low.

A number of cross-sectional serologic studies addressed the probability of exposure to multiple tick-borne pathogens. The frequency of exposure to both *B. burgdorferi* and *A. phagocytophilum* estimated by serologic methods usually does not exceed 1-5% [12, 19, 22, 28, 40, 46, 62, 74, 95, 114]. This includes of course both true coinfections and sequential infections since most serologic studies could not distinguish between the two scenarios.

One prospective seroincidence study was conducted among 671 persons with high risk exposures in a region of New York where both Lyme disease and human granulocytic anaplasmosis (ehrlichiosis) are endemic [48]. None of the participants seroconverted to two or more agents during 1 year of follow-up. The absence of coincident seroconversions suggests that the absolute risk of coinfection is low even among persons with frequent outdoor exposures in endemic areas.

Thus, the number of confirmed cases of simultaneous infections with *B. burgdorferi* and *A. phagocytophilum* in humans seems lower than the potential number of coinfection cases that could be expected based on the frequency of coinfection in ticks (2-26%) plus the frequency of multiple tick bites. This incongruence may be due to symptoms of one infection masking the other; the inability of co-infected ticks to transmit all of the pathogens simultaneously; or the failure of one (second?) pathogen to establish itself and cause a clinical illness in the presence of another infection. According to the studies described in this chapter, ticks simultaneously infected with *B. burgdorferi* and *A. phagocytophilum* can successfully transmit both agents to susceptible hosts, but mice already infected with one of those agents become less susceptible to the other one. It is not known whether humans may still develop a detectable immune response to an unapparent infection without succumbing to a clinical illness, or whether an unapparent infection may become acute if/when a patient is treated against the "primary" infection but not against a "secondary" one. These questions are yet to be answered in carefully controlled studies.

Conclusion

There was no evidence of interaction between the agents of Lyme disease and HGE in the vector tick *I. scapularis*. Presence of either agent in ticks did not interfere with acquisition of the other agent from an infected host. Transmission of the agents of Lyme disease and HGE by individual ticks was equally efficient and independent. Dually infected ticks transmitted each pathogen to susceptible hosts as efficiently as ticks infected with one pathogen only, and most dually infected ticks were able to transmit both pathogens to a susceptible host.

On the other hand, naive white-footed mice are apparently susceptible to coinfection when the two agents are introduced simultaneously. Interference between *A. phagocytophilum* and *B. burgdorferi* occurs within a natural vertebrate reservoir host as an outcome of sequential coinfection. Suppressive effects on transmission of the second agent are probably due to the immunological status developing after the infection with the first agent. Detailed immunological and microbiological studies are necessary to uncover particular mechanisms of direct or indirect interaction between these pathogens in vertebrate hosts including humans.

References

[1] Ackermann, R. 1983. [The spirochetal etiology of erythema chronicum migrans and of meningo-polyneuritis Garin-Bujadoux-Bannwarth] [German]. *Z. Hautkr.* 58(22): 1616-1621.

[2] Ahkee, S., and J. Ramirez. 1996. A case of concurrent Lyme meningitis with ehrlichiosis. *Scand. J. Infect. Dis.* 28(5): 527-8.

[3] Aristova, V. A., and N. M. Okulova. 1976. [Influence of the mobility of small forest mammals on their infestation with larvae and nymphs of ixodid ticks] [Russian]. *Fauna. Ecol. Gryzunov* 13: 88-100.

[4] Bakken, J. S., and J. S. Dumler. 2000. Human granulocytic ehrlichiosis. *Clin. Infect. Dis.* 31(2): 554-560.

[5] Bakken, J. S., J. Krueth, R. L. Tilden, J. S. Dumler, and B. E. Kristiansen. 1996. Serological evidence of human granulocytic ehrlichiosis in Norway. *J. Clin. Microbiol.* 15(10): 829-832.

[6] Barlough, J. E., J. E. Madigan, E. DeRock, J. S. Dumler, and J. S. Bakken. 1995. Protection against *Ehrlichia equi* is conferred by prior infection with the human granulocytotropic *Ehrlichia* (HGE agent). *J. Clin. Microbiol.* 33(12): 3333-3334.

[7] Barton, L. L., A. Luisiri, J. E. Dawson, G. W. Letson, and T. J. Quan. 1990. Simultaneous infection with an *Ehrlichia* and *Borrelia burgdorferi* in a child. *Ann. N. Y. Acad. Sci.* 590: 68-9.

[8] Batungbacal, M. R., and G. R. Scott. 1982. Suppression of the immune response to clostridial vaccine by tick-borne fever. *J. Comp. Pathol.* 92(3): 409-413.

[9] Batungbacal, M. R., and G. R. Scott. 1982. Tick-borne fever and concurrent parainfluenza-3 virus infection in sheep. *J. Comp. Pathol.* 92(3): 415-428.

[10] Baumgarten, B. U., M. Rollinghoff, and C. Bogdan. 1999. Prevalence of *Borrelia burgdorferi* and granulocytic and monocytic Ehrlichiae in *Ixodes ricinus* ticks from southern Germany. *J. Clin. Microbiol.* 37(11): 3448-3451.

[11] Belongia, E. A., K. D. Reed, P. D. Mitchell, P. H. Chyou, N. Mueller-Rizner, M. F. Finkel, and M. E. Schriefer. 1999. Clinical and epidemiological features of early Lyme disease and human granulocytic ehrlichiosis in Wisconsin. *Clin. Infect. Dis.* 29(6): 1472-1477.

[12] Björsdorff, A., B. Wittesjo, J. Berglund, R. F. Massung, and I. Eliasson. 2002. Human granulocytic ehrlichiosis as a common cause of tick-associated fever in Southeast Sweden: Report from a prospective clinical study. *Scand. J. Infect. Dis.* 34(3): 187-191.

[13] Brodie, T. A., P. H. Holmes, and G. M. Urquhart. 1986. Some aspects of tick-borne diseases of British sheep. *Vet. Rec.* 118(15): 415-418.

[14] Brouqui, P., J. S. Dumler, R. Lienhard, M. Brossard, and D. Raoult. 1995. Human granulocytic ehrlichiosis in Europe [letter]. *Lancet* 346(8977): 782-783.

[15] Brunet, L. R., A. Spielman, and S. R. Telford, 3rd. 1995. Density of Lyme disease spirochetes within deer ticks collected from zoonotic sites (Short report). *Am. J. Trop. Med. Hyg.* 53(3): 300-302.

[16] Brun-Hansen, H., D. A. Christensson, F. Hardeng, and H. Gronstol. 1997. Experimental infection with *Ehrlichia phagocytophila* and *Babesia divergens* in cattle. *Zentralbl. Veterinarmed* [B]. 44(4): 235-243.

[17] Chen, S. M., J. S. Dumler, J. S. Bakken, and D. H. Walker. 1994. Identification of a granulocytotropic *Ehrlichia* species as the etiologic agent of human disease. *J. Clin. Microbiol.* 32(3): 589-595.

[18] Christova, I., L. Schouls, I. van de Pol, J. Park, S. Panayotov, V. Lefterova, T. Kantardjiev, and J. S. Dumler. 2001. High prevalence of granulocytic Ehrlichiae and *Borrelia burgdorferi sensu lato* in *Ixodes ricinus* ticks from Bulgaria. *J. Clin. Microbiol.* 39(11): 4172-4174.

[19] Christova, I. S., and J. S. Dumler. 1999. Human granulocytic ehrlichiosis in Bulgaria. *Am. J. Trop. Med. Hyg.* 60(1): 58-61.

[20] Cinco, M., D. Padovan, R. Murgia, M. Heldtander, and E. O. Engvall. 1998. Detection of HGE agent-like *Ehrlichia* in *Ixodes ricinus* ticks in northern Italy by PCR. *Wien. Klin. Wochenschr.* 110(24): 898-900.

[21] Daniels, T. J., R. C. Falco, I. Schwartz, S. Varde, and R. G. Robbins. 1997. Deer ticks (*Ixodes scapularis*) and the agents of Lyme disease and human granulocytic ehrlichiosis in a New York City park. *Emerg. Infect. Dis.* 3(3): 353-355.

[22] De Martino, S. J., J. A. Carlyon, and E. Fikrig. 2001. Coinfection with *Borrelia burgdorferi* and the agent of human granulocytic ehrlichiosis. *N. Engl. J. Med.* 345(2): 150-151.

[23] Des Vignes, F., and D. Fish. 1997. Transmission of the agent of human granulocytic ehrlichiosis by host-seeking *Ixodes scapularis* (Acari:Ixodidae) in southern New York state. *J. Med. Entomol.* 34(4): 379-382.

[24] Des Vignes, F., M. L. Levin, and D. Fish. 1999. Comparative vector competence of *Dermacentor variabilis* and *Ixodes scapularis* (Acari: Ixodidae) for the agent of human granulocytic ehrlichiosis. *J. Med. Entomol.* 36(2): 182-185.

[25] Duffy, J., M. R. Pittlekow, C. P. Kolbert, B. J. Rutledge, and D. H. Persing. 1997. Coinfection with *Borrelia burgdorferi* and the agent of human granulocytic ehrlichiosis. *Lancet* 349(9049): 399.

[26] Dumler, J. S., K. M. Asanovich, J. S. Bakken, P. Richter, R. Kimsey, and J. E. Madigan. 1995. Serologic cross-reactions among *Ehrlichia equi*, *Ehrlichia phagocytophila*, and human granulocytic ehrlichia. *J. Clin. Microbiol.* 33(5): 1098-1103.

[27] Dumler, J. S., A. F. Barbet, C. P. J. Bekker, G. A. Dasch, G. H. Palmer, S. C. Ray, Y. Rikihisa, and F. R. Rurangirwa. 2001. Reorganization of genera in the families Rickettsiaceae and Anaplasmataceae in the order Rickettsiales: unification of some species of *Ehrlichia* with *Anaplasma*, *Cowdria* with *Ehrlichia* and *Ehrlichia* with *Neorickettsia*, descriptions of six new species combinations and designation of *Ehrlichia equi* and 'HGE agent' as subjective synonyms of *Ehrlichia phagocytophila*. *Int. J. Syst. Evol. Microbiol.* 51(6): 2145-2165.

[28] Dumler, J. S., L. Dotevall, R. Gustafson, and M. Granstrom. 1997. A population-based seroepidemiologic study of human granulocytic ehrlichiosis and Lyme borreliosis on the west coast of Sweden. *J. Infect. Dis.* 175(3): 720-722.

[29] Dunaeva, T. N. 1970. [Characteristics of tularemia in blue hares in mixed infections] [Russian]. *Zool. Zhurn.* 49(12): 1864-1868.

[30] Dunaeva, T. N. 1971. [Mixed infection in experimental tularemia in rabbits] [Russian]. *Zh. Mikrobiol. Epidemiol. Immunobiol.* 48(8): 125-30.

[31] Dunaeva, T. N. 1972. [Peculiarities of relationships between warm-blooded animals and pathogens in mixed infections] [Russian], p. 88-107. *In* P. A. Petrischeva (ed.), Itogi razvitiia ucheniia o prirodnoi ochagovosti boleznei cheloveka i dal'neishie zadachi. *Meditsina,* Moscow.

[32] Dunaeva, T. N., E. V. Ananova, and K. N. Shlygina. 1978. [Mechanism of the changes in the animal reactivity to tularemia in mixed infection] [Russian]. *Zh. Mikrobiol. Epidemiol. Immunobiol.*(1): 57-61.

[33] Dunaeva, T. N., and K. N. Shlygina. 1977. [Change in phagocytic activity toward the agent of tularemia in highly sensitive animals with mixed infections] [Russian]. *Zh. Mikrobiol. Epidemiol. Immunobiol.*(4): 86-91.

[34] Fingerle, V., J. L. Goodman, R. C. Johnson, T. J. Kurtti, U. G. Munderloh, and B. Wilske. 1997. Human granulocytic ehrlichiosis in southern Germany: increased seroprevalence in high-risk groups. *J. Clin. Microbiol.* 35(12): 3244-3247.

[35] Fingerle, V., U. G. Munderloh, G. Liegl, and B. Wilske. 1999. Coexistence of ehrlichiae of the phagocytophila group with *Borrelia burgdorferi* in *Ixodes ricinus* from Southern Germany. *Med. Microbiol. Immunol.* (Berl) 188(3): 145-149.

[36] Foggie, A. 1951. Studies on the infectious agent of tick-borne fever in sheep. *J. Pathol. Bacteriol.* 63(1): 1-15.

[37] Foster, W. N., and A. E. Cameron. 1970. Observations on the functional integrity of neutrophil leucocytes infected with tick-borne fever. *J. Comp. Pathol.* 80(3): 487-491.

[38] Foster, W. N., and A. E. Cameron. 1968. Thrombocytopenia in sheep associated with experimental tick- borne fever infection. *J. Comp. Pathol.* 78(2): 251-254.

[39] Fraser, C. M., S. Casjens, W. M. Huang, G. G. Sutton, R. Clayton, R. Lathigra, O. White, K. A. Ketchum, R. Dodson, E. K. Hickey, M. Gwinn, B. Dougherty, J. F. Tomb, R. D. Fleischmann, D. Richardson, J. Peterson, A. R. Kerlavage, J. Quackenbush, S. Salzberg, M. Hanson, R. van Vugt, N. Palmer, M. D. Adams, J. Gocayne, and J. C. Venter. 1997. Genomic sequence of a Lyme disease spirochaete, *Borrelia burgdorferi*. *Nature* 390(6660): 580-586.

[40] Fritz, C. L., A. M. Kjemtrup, P. A. Conrad, G. R. Flores, G. L. Campbell, M. E. Schriefer, D. Gallo, and D. J. Vugia. 1997. Seroepidemiology of emerging tickborne infectious diseases in a Northern California community. *J. Infect. Dis.* 175(6): 1432-1439.

[41] Gern, L., and O. Rais. 1996. Efficient transmission of *Borrelia burgdorferi* between cofeeding *Ixodes ricinus* ticks (Acari: Ixodidae). *J. Med. Entomol.* 33(1): 189-192.

[42] Gilmour, J. L., T. A. Brodie, and P. H. Holmes. 1982. Tick-borne fever and pasteurellosis in sheep [letter]. *Vet. Rec.* 111(22): 512.

[43] Gokce, H. I., and Z. Woldehiwet. 1999. The effects of *Ehrlichia* (*Cytoecetes*) *phagocytophila* on the clinical chemistry of sheep and goats. *Zentralbl Veterinarmed* [B]. 46(2): 93-103.

[44] Gokce, H. I., and Z. Woldehiwet. 1999. *Ehrlichia (Cytoecetes) phagocytophila* predisposes to severe contagious ecthyma (Orf) in lambs. *J. Comp. Pathol.* 121(3): 227-40.

[45] Gordon, W. S., A. Brownlee, D. R. Wilson, and J. MacLeod. 1932. "Tick-borne fever" (a hitherto undescribed disease of sheep). *J. Comp. Pathol. Ther* 45: 301-312.

[46] Guillaume, B., P. Heyman, S. Lafontaine, C. Vandenvelde, M. Delmee, and G. Bigaignon. 2002. Seroprevalence of human granulocytic ehrlichlosis infection in Belgium. *Eur. J. Clin. Microbiol. Infect. Dis.* 21(5): 397-400.

[47] Hildebrandt, A., K. H. Schmidt, B. Wilske, W. Dorn, E. Straube, and V. Fingerle. 2003. Prevalence of four species of *Borrelia burgdorferi* sensu lato and coinfection

[ ] with *Anaplasma phagocytophila* in *Ixodes ricinus* ticks in central Germany. *Eur. J. Clin. Microbiol. Infect. Dis.* 22(6): 364-367.

[48] Hilton, E., J. DeVoti, J. L. Benach, M. L. Halluska, D. J. White, H. Paxton, and J. S. Dumler. 1999. Seroprevalence and seroconversion for tickborne diseases in a high-risk population in the northeast United States. *Am. J. Med* 106(4): 404-409.

[49] Hofmeister, E. K., J. Magera, L. Sloan, C. Kolbert, J. Hanson, and D. H. Persing. 1996. *Borrelia burgdorferi* proteins are recognized by antibodies from nice experimentally infected with the agent of human granulocytic ehrlichiosis. *Presented at the* VII International Congress on Lyme Borreliosis. June 16-21, 1996., San Francisco, CA.

[50] Holden, K., J. T. Boothby, S. Anand, and R. F. Massung. 2003. Detection of *Borrelia burgdorferi, Ehrlichia chaffeensis,* and *Anaplasma phagocytophilum* in ticks (Acari: Ixodidae) from a coastal region of California. *J. Med. Entomol.* 40(4): 534-539.

[51] Johnson, R. C., J. Goodman, S. M. Engstrom, L. A. Coleman, C. C. Kodner, and M. D. Ravin. 1996. Human granulocytic ehrlichiosis: Evolution of the humoral response and coinfection with *Borrelia burgdorferi* in the United States of America. *Presented at the* International ScientificConference: "Tick-borne viral, rickettsial and bacterial infections", Irkutsk, RU.

[52] Korenberg, E. I., G. G. Bannova, Y. V. Kovalevskii, and A. S. Karavanov. 1988. [Intrapopulation differences in the infectivity of adult *Ixodes persulcatus* P. Sch. with the tick-borne encephalitis virus and an assessment of its total content in ticks] [Russian]. *Vopr. Virusol.* 33(4): 456-61.

[53] Korenberg, E. I., and Y. V. Kovalevskii. 1995. Variation in parameters affecting risk of human disease due to TBE virus. *Folia Parasitol.* (Praha). 42(4): 307-312.

[54] Kovalevskii, Y. V., and E. I. Korenberg. 1995. Differences in *Borrelia infections* in adult *Ixodes persulcatus* and *Ixodes ricinus* ticks (Acari: Ixodidae) in populations of north-western Russia. *Exp. Appl. Acarol.* 19(1): 19-29.

[55] Kovalevskii, Y. V., and E. I. Korenberg. 1990. [Factors that determine the possibility of tick-borne encephalitis infection. 3. The probability of human contact with an infected vector in the central taiga forests of Khabarovsk Territory] [Russian]. *Med. Parazitol. Parazit. Bolezn* 59(3): 5-8.

[56] Kovalevskii, Y. V., E. I. Korenberg, T. V. Kutlina, and O. A. Ustinova. 1996. [Standards for reviewing preparations in the study of ixodid tick nymphs by dark-field microscopy in borreliosis foci] [Russian]. *Med. Parazitol. Parazit. Bolezn* 65(4): 18-21.

[57] Kovalevskii, Y. V., E. I. Korenberg, and I. G. Nikitochkin. 1991. [The optimization of a method for assessing the contagiosity and the degree of individual infectiousness of ticks with *Borrelia*] [Russian]. Med. Parazitol. *Parazit. Bolezn* 60(3): 18-21.

[58] Krause, P. J., K. McKay, C. A. Thompson, V. K. Sikand, R. Lentz, T. Lepore, L. Closter, D. Christianson, S. R. Telford, D. Persing, J. D. Radolf, and A. Spielman. 2002. Disease-specific diagnosis of coinfecting tickborne zoonoses: Babesiosis, human granulocytic ehrlichiosis, and Lyme disease. *Clin. Infect. Dis.* 34(9): 1184-1191.

[59] Lane, R. S., J. E. Foley, L. Eisen, E. T. Lennette, and M. A. Peot. 2001. Acarologic risk of exposure to emerging tick-borne bacterial pathogens in a semirural community in northern California. *Vector borne zoonotic dis.* 1(3): 197-210.

[60] Larsen, H. J., G. Overnes, H. Waldeland, and G. M. Johansen. 1994. Immunosuppression in sheep experimentally infected with *Ehrlichia phagocytophila*. *Res. Vet. Sci.* 56(2): 216-224.

[61] Leutenegger, C. M., N. Pusterla, C. N. Mislin, R. Weber, and H. Lutz. 1999. Molecular evidence of coinfection of ticks with *Borrelia burgdorferi* sensu lato and the human granulocytic ehrlichiosis agent in Switzerland. *J. Clin. Microbiol.* 37(10): 3390-3391.

[62] Levin, A. E., and T. W. Talaska. 1999. Serologic evidence of coinfection with HGE and *Borrelia burgdorferi* in Germany. *Int. J. Med. Microbiol.* 289(5-7): 768-769.

[63] Levin, M., J. F. Levine, C. S. Apperson, D. E. Norris, and P. B. Howard. 1995. Reservoir competence of the rice rat (Rodentia: Cricetidae) for *Borrelia burgdorferi*. *J. Med. Entomol.* 32(2): 138-142.

[64] Levin, M., M. Papero, and D. Fish. 1997. Feeding density influences acquisition of *Borrelia burgdorferi* in larval *Ixodes scapularis* (Acari: Ixodidae). *J. Med. Entomol.* 34(5): 569-572.

[65] Levin, M. L., F. des Vignes, and D. Fish. 1999. Disparity in the natural cycles of *Borrelia burgdorferi* and the agent of human granulocytic ehrlichiosis. *Emerg. Infect. Dis.* 5(2): 204-208.

[66] Levin, M. L., Y. V. Kovalevskii, A. Y. Piskunova, and T. V. Schegoleva. 1993. Evaluation of individual ticks infection rate with Lyme disease agent by microscopic examination of fixed smears, p. 157-162. In E. I. Korenberg (ed.), *Problems of Tick-Borne Borrelioses*. Academy of Medical Scienses of Russia, Moscow.

[67] Levin, M. L., W. L. Nicholson, R. Massung, J. W. Sumner, and D. Fish. 2002. Comparison of the reservoir competence of medium-sized mammals and *Peromyscus leucopus* for *Anaplasma phagocytophila* in Connecticut. *Vector borne zoonotic dis.* 2(3): 125-136.

[68] Lotric-Furlan, S., M. Petrovec, T. Avsic-Zupanc, W. L. Nicholson, J. W. Sumner, J. E. Childs, and F. Strle. 1998. Human ehrlichiosis in Central Europe. *Wien. Klin. Wochenschr.* 110(24): 894-897.

[69] MacLeod, J. R., and W. S. Gordon. 1933. Studies of tick-borne fever of sheep. 1. Transmission by the tick *Ixodes ricinus*, with a description of the disease produced. *Parasitology* 25: 273-285.

[70] Magnarelli, L. A., J. F. Anderson, K. C. Stafford, 3d, and J. S. Dumler. 1997. Antibodies to multiple tick-borne pathogens of babesiosis, ehrlichiosis, and Lyme disease in white-footed mice. *J. Wildl. Dis.* 33(3): 466-473.

[71] Magnarelli, L. A., K. C. Stafford, 3rd, T. N. Mather, M. T. Yeh, K. D. Horn, and J. S. Dumler. 1995. Hemocytic rickettsia-like organisms in ticks: serologic reactivity with antisera to ehrlichiae and detection of DNA of agent of human granulocytic ehrlichiosis by PCR. *J. Clin. Microbiol.* 33(10): 2710-2714.

[72] Mather, T. N., S. R. Telford, 3d, S. I. Moore, and A. Spielman. 1990. *Borrelia burgdorferi* and *Babesia microti*: efficiency of transmission from reservoirs to vector ticks (*Ixodes dammini*). *Exp. Parasitol.* 70(1): 55-61.

[73] Mazzella, F. M., A. Roman, and A. Perez. 1996. A case of concurrent presentation of human ehrlichiosis and Lyme disease in Connecticut. *Conn. Med.* 60(9): 515-9.

[74] Mitchell, P. D., K. D. Reed, and J. M. Hofkes. 1996. Immunoserologic evidence of coinfection with *Borrelia burgdorferi*, *Babesia microti*, and human granulocytic *Ehrlichia* species in residents of Wisconsin and Minnesota. *J. Clin. Microbiol.* 34(3): 724-727.

[75] Moss, W. J., and J. S. Dumler. 2003. Simultaneous infection with *Borrelia burgdorferi* and human granulocytic ehrlichiosis. *Pediatr. Infect. Dis. J.* 22(1): 91-92.

[76] Munro, R., A. R. Hunter, G. MacKenzie, and D. A. McMartin. 1982. Pulmonary lesions in sheep following experimental infection by *Ehrlichia phagocytophilia* and *Chlamydia psittaci*. *J. Comp. Pathol.* 92(1): 117-129.

[77] Nadelman, R. B., H. W. Horowitz, T. C. Hsieh, J. M. Wu, M. E. Aguero-Rosenfeld, I. Schwartz, J. Nowakowski, S. Varde, and G. P. Wormser. 1997. Simultaneous human granulocytic ehrlichiosis and Lyme borreliosis. *N. Engl. J. Med.* 337(1): 27-30.

[78] Nadelman, R. B., and G. P. Wormser. 1998. Management of tick bites and early Lyme disease, p. 49-76. *In* D. Rahn and J. Evans (ed.), *Lyme disease*. American college of Physicians, Philadelphia.

[79] Nakayama, Y., and A. Spielman. 1989. Ingestion of Lyme disease spirochetes by ticks feeding on infected hosts [letter]. *J. Infect. Dis.* 160(1): 166-167.

[80] Nicholson, W. L., J. A. Comer, J. W. Sumner, C. Gingrich-Baker, R. T. Coughlin, L. A. Magnarelli, J. G. Olson, and J. E. Childs. 1997. An indirect immunofluorescence assay using a cell culture- derived antigen for detection of antibodies to the agent of human granulocytic ehrlichiosis. *J. Clin. Microbiol.* 35(6): 1510-1516.

[81] Nicholson, W. L., S. Muir, J. W. Sumner, and J. E. Childs. 1998. Serologic evidence of infection with *Ehrlichia* spp. in wild rodents (Muridae: Sigmodontinae) in the United States. *J. Clin. Microbiol.* 36(3): 695-700.

[82] Nuttall, P. A. 1998. Displaced tick-parasite interactions at the host interface. *Parasitology* 116: S65-S72.

[83] Ogden, N. H., P. A. Nuttall, and S. E. Randolph. 1997. Natural Lyme disease cycles maintained via sheep by cofeeding ticks. *Parasitology* 115(6): 591-599.

[84] Overas, J., A. Lund, M. J. Ulvund, and H. Waldeland. 1993. Tick-borne fever as a possible predisposing factor in septicaemic pasteurellosis in lambs. *Vet. Rec.* 133(16): 398.

[85] Pancholi, P., C. P. Kolbert, P. D. Mitchell, K. D. Reed, J. S. Dumler, J. S. Bakken, S. R. Telford, and D. H. Persing. 1995. *Ixodes dammini* as a potential vector of human granulocytic ehrlichiosis. *J. Infect. Dis.* 172(4): 1007-1012.

[86] Patrican, L. A. 1997. Acquisition of Lyme disease spirochetes by cofeeding *Ixodes scapularis* ticks. *Am. J. Trop. Med. Hyg.* 57(5): 589-593.

[87] Pavlovskii, E. N. 1937. [The doctrine of biocenoses in application to some parasitological problems] [Russian]. *Izv. Akad. Nauk. Ser. Biol.*(4): 1385-1422.

[88] Persing, D. H. 1997. The cold zone: A curious convergence of tick-transmitted diseases. *Clin. Infect. Dis.* 25(Suppl:1): S 35-S 42.

[89] Piesman, J., T. C. Hicks, R. J. Sinsky, and G. Obiri. 1987. Simultaneous transmission of *Borrelia burgdorferi* and *Babesia microti* by individual nymphal *Ixodes dammini* ticks. *J. Clin. Microbiol.* 25(10): 2012-2013.

[90] Piesman, J., T. N. Mather, R. J. Sinsky, and A. Spielman. 1987. Duration of tick attachment and *Borrelia burgdorferi* transmission. *J. Clin. Microbiol.* 25(3): 557-558.

[91] Piesman, J., G. O. Maupin, E. G. Campos, and C. M. Happ. 1991. Duration of adult female *Ixodes dammini* attachment and transmission of *Borrelia burgdorferi*, with description of a needle aspiration isolation method. *J. Infect. Dis.* 163(4): 895-897.

[92] Pomanskaia, L. A. 1970. [Interrelations between the infective agents of tularemia and listeriosis in vitro and in vivo] [Russian]. *Zh. Mikrobiol. Epidemiol. Immunobiol.* 47(4): 36-42.

[93] Purnell, R. E., D. W. Brocklesby, D. J. Hendry, and E. R. Young. 1976. Separation and recombination of *Babesia divergens* and *Ehrlichia phagocytophila* from a field case of redwater from Eire. *Vet. Rec.* 99(21): 415-417.

[94] Purnell, R. E., E. R. Young, D. W. Brocklesby, and D. J. Hendry. 1977. The haematology of experimentally-induced *B. divergens* and *E. phagocytophila* infections in splenectomised calves. *Vet. Rec.* 100(1): 4-6.

[95] Pusterla, N., R. Weber, C. Wolfensberger, G. Schar, R. Zbinden, W. Fierz, J. E. Madigan, J. S. Dumler, and H. Lutz. 1998. Serological evidence of human granulocytic ehrlichiosis in Switzerland. *J. Clin. Microbiol.* 17(3): 207-209.

[96] Reed, K. D., P. D. Mitchell, D. H. Persing, C. P. Kolbert, and V. Cameron. 1995. Transmission of human granulocytic ehrlichiosis [letter; comment]. *JAMA* 273(1): 23.

[97] Reid, H. W., D. Buxton, I. Pow, T. A. Brodie, P. H. Holmes, and G. M. Urquhart. 1986. Response of sheep to experimental concurrent infection with tick-borne fever (*Cytoecetes phagocytophila*) and louping-ill virus. *Res. Vet. Sci.* 41(1): 56-62.

[98] Rikihisa, Y. 1991. The tribe Ehrlichieae and ehrlichial diseases. *Clin. Microbiol. Rev.* 4(3): 286-308.

[99] Schauber, E. M., S. J. Gertz, W. T. Maple, and R. S. Ostfeld. 1998. Coinfection of blacklegged ticks (Acari: Ixodidae) in Dutchess County, New York, with the agents of Lyme disease and human granulocytic ehrlichiosis. *J. Med. Entomol.* 35(5): 901-3.

[100] Schouls, L. M., I. Van de Pol, S. G. T. Rijpkema, and C. S. Schot. 1999. Detection and identification of *Ehrlichia*, *Borrelia burgdorferi* sensu lato, and *Bartonella* species in Dutch *Ixodes ricinus* ticks. *J. Clin. Microbiol.* 37(7): 2215-2222.

[101] Schwartz, I., D. Fish, and T. J. Daniels. 1997. Prevalence of the rickettsial agent of human granulocytic ehrlichiosis in ticks from a hyperendemic focus of Lyme disease. *N. Engl. J. Med.* 337(1): 49-50.

[102] Schwartz, I., S. Varde, R. B. Nadelman, G. P. Wormser, and D. Fish. 1997. Inhibition of efficient polymerase chain reaction amplification of *Borrelia burgdorferi* DNA in blood-fed ticks. *Am. J. Trop. Med. Hyg.* 56(3): 339-342.

[103] Sinsky, R. J., and J. Piesman. 1989. Ear punch biopsy method for detection and isolation of *Borrelia burgdorferi* from rodents. *J. Clin. Microbiol.* 27(8): 1723-1727.

[104] Skotarczak, B., A. Rymaszewska, B. Wodecka, and M. Sawczuk. 2003. Molecular evidence of coinfection of *Borrelia burgdorferi* sensu lato, human granulocytic ehrlichiosis agent, and *Babesia microti* in ticks from northwestern Poland. *J. Parasitol.* 89(1): 194-196.

[105] Stanczak, J., M. Racewicz, W. Kruminis-Lozowska, and B. Kubica-Biernat. 2002. Coinfection of *Ixodes ricinus* (Acari: Ixodidae) in northern Poland with the agents of Lyme borreliosis (LB) and human granulocytic ehrlichiosis (HGE). *Int. J. Med. Microbiol.* 291(Suppl 33): 198-201.

[106] Stannard, A. A., D. H. Gribble, and R. S. Smith. 1969. Equine ehrlichiosis: a disease with similarities to tick-borne fever and bovine petechial fever. *Vet. Rec.* 84(6): 149-150.

[107] Steere, A. C., G. McHugh, C. Suarez, J. Hoitt, N. Damle, and V. K. Sikand. 2003. Prospective study of coinfection in patients with erythema migrans. *Clin. Infect. Dis.* 36(8): 1078-1081.

[108] Sumner, J. W., W. L. Nicholson, and R. F. Massung. 1997. PCR amplification and comparison of nucleotide sequences from the *groESL* heat shock operon of *Ehrlichia* species. *J. Clin. Microbiol.* 35(8): 2087-2092.

[109] Taylor, A. W., H. H. Holman, and W. S. Gordon. 1941. Attempts to reproduce the pyaemia associated with tick-bite. *Vet. Rec.* 53: 339-344.

[110] Telford, S. R., 3rd, J. E. Dawson, P. Katavolos, C. K. Warner, C. P. Kolbert, and D. H. Persing. 1996. Perpetuation of the agent of human granulocytic ehrlichiosis in a deer tick-rodent cycle. *Proc. Natl. Acad. Sci. U. S. A.* 93(12): 6209-6214.

[111] Thomas, V., J. Anguita, S. W. Barthold, and E. Fikrig. 2001. Coinfection with *Borrelia burgdorferi* and the agent of human granulocytic ehrlichiosis alters murine immune responses, pathogen burden, and severity of Lyme arthritis. *Infect. Immun.* 69(5): 3359-3371.

[112] Varde, S., J. Beckley, and I. Schwartz. 1998. Prevalence of tick-borne pathogens in *Ixodes scapularis* in a rural New Jersey County. *Emerg. Infect. Dis.* 4(1): 97-99.

[113] Walker, D. H., and J. S. Dumler. 1996. Emergence of the ehrlichioses as human health problems. *Emerg. Infect. Dis.* 2(1): 18-29.

[114] Weber, R., N. Pusterla, M. Loy, C. M. Leutenegger, G. Schar, D. Baumann, C. Wolfensberger, and H. Lutz. 2000. [Serologic and clinical evidence for endemic occurrences of human granulocytic ehrlichiosis in North-Eastern Switzerland] [German]. *Schweiz. Med. Wochenschr.* 130(41): 1462-70.

[115] Woldehiwet, Z. 1987. Depression of lymphocyte response to mitogens in sheep infected with tick-borne fever. *J. Comp. Pathol.* 97(6): 637-643.

[116] Woldehiwet, Z. 1987. The effects of tick-borne fever on some functions of polymorphonuclear cells of sheep. *J. Comp. Pathol.* 97(4): 481-485.

[117] Woldehiwet, Z. 1991. Lymphocyte subpopulations in peripheral blood of sheep experimentally infected with tick-borne fever. *Res. Vet. Sci.* 51(1): 40-43.

[118] Woldehiwet, Z., and G. R. Scott. 1982. Tick-borne fever: leucocyte migration inhibition. *Vet. Microbiol.* 7(5): 437-445.

[119] Wormser, G. P., H. W. Horowitz, J. S. Dumler, I. Schwartz, and M. Aguero-Rosenfeld. 1996. False-positive Lyme disease serology in human granulocytic ehrlichiosis [letter]. *Lancet* 347(9006): 981-982.

[120] Wormser, G. P., H. W. Horowitz, J. Nowakowski, D. Mckenna, J. S. Dumler, S. Varde, I. Schwartz, C. Carbonaro, and M. Aguero-Rosenfeld. 1997. Positive Lyme disease serology in patients with clinical and laboratory evidence of human granulocytic ehrlichiosis. *Am. J. Clin. Pathol.* 107(2): 142-147.

[121] Zeidner, N. S., T. R. Burkot, R. Massung, W. L. Nicholson, M. C. Dolan, J. S. Rutherford, B. J. Biggerstaff, and G. O. Maupin. 2000. Transmission of the agent of human granulocytic ehrlichiosis by *Ixodes spinipalpis* ticks: Evidence of an enzootic cycle of dual infection with *Borrelia burgdorferi* in northern Colorado. *J. Infect. Dis.* 182(2): 616-619.

Short Communication A

Early Detection of Baculovirus Expression and Infection in Lepidopteran Larvae Fed Occlusion Bodies of an AcMNPV Recombinant Carrying a Red Fluorescent Protein Gene[*]

Arthur H. McIntosh[†] and James J. Grasela

USDA, Agricultural Research Service, Biological Control of Insects Research Laboratory, 1503 S. Providence Road, Research Park
Columbia, Missouri 65203-3535, USA

Abstract

A method was devised utilizing a baculovirus recombinant (AcMNPV hsp70Red) carrying a red fluorescent protein (RFP) gene under the early heat shock promoter (hsp70) to assess potential infectivity of larvae fed occlusion bodies. A time study was employed whereby first and third instars of *Trichoplusia ni*, *Heliothis subflexa* and *Helicoverpa zea* could be followed for red fluorescence under a UV-light inverted microscope at various time intervals following diet surface feeding of recombinant occlusion bodies. Larvae of *T. ni* and *H. subflexa* that are permissive to AcMNPV showed 100 % fluorescence by 24h and 36h respectively for 1st instars exposed to 1×10^5

---

[*] Mention of trade names or commercial product in the publication is solely for the purpose of providing specific information and does not imply recommendation or endorsement by the U. S. Department of Agriculture.
All programs and services of the U. S. Department of Agriculture are offered on a nondiscriminatory basis without regard to race, color, national origin, religion, sex, age, marital status, or handicap.
[†] Corresponding author: Arthur H. McIntosh, USDA, ARS, BCIRL; 1503 S. Providence Rd. Columbia, MO 65203-3535; Ph. (573) 875-5361; Fax. (573) 875-5364; Email:mcintosha@missouri.edu

OB. Even when the concentration of OB was reduced 100 fold to 1×10^3 OB, 94%-100% fluorescence was observed by 24h post feeding. *H. zea* which is less permissive to AcMNPV than *T. ni* and *H. subflexa* and when 1st instars were fed 1×10^5 OB showed 100% red fluorescence at 48h post exposure as contrasted with 86%-90% fluorescence at the 96h exposure period for the lower dose of 1×10^3 OB. *T. ni* and *H. subflexa* third instars fed 1×10^5 OB gave 100% fluorescence between 24h (*T.ni*) and 36h- 72h (*H. subflexa*) post exposure whereas at the lower dose of 1×10^3 OB 100% fluorescence was achieved between 24h-72h post exposure. Third instar *H. zea* larvae fed both doses did not attain 100% fluorescence. All larvae of each species that gave 100% fluorescence resulted in 100% mortality. Fluorescent fifty (FL_{50}) values, the time taken to attain 50% fluorescence of exposed larvae, were calculated for all categories. *T. ni* gave the shortest times for both 1st ($FL_{50} = 12.8h$) and 3rd instar ($FL_{50} = 16.0h$) at the OB dose of 1×10^5 followed by *H. subflexa* and *H. zea*. The same pattern was observed at the lower OB dose of 1×10^3.

Introduction

Since their inception as biopesticidal agents for the control of lepidopterous pests of forest, field, fruit and vegetable crops, baculoviruses have found many applications and uses such as in molecular biology for the development of baculovirus expression vectors (BEV); in the engineering of insect toxin genes for the production of faster acting baculoviruses; for the production of many thousands of recombinant proteins and more recently as gene delivery vectors in the medical field (Luckow and Summers, 1988; Hammock and Fowler, 1991; O'Reilly et al., 1992; Maeda et. al., 1991; McCutchen et. al., 1991; Tomalski and Miller, 1991; Boyce and Bucher, 1996; Kost et. al., 2005).

Baculoviruses are classified in the family Baculoviridae and are comprised of two genera, the Nucleopolyhedrovirus and the Granulovirus (Fauquet et al. 2005) but a new classification has been recently proposed (Jehle et al., 2006). Baculoviruses are large double stranded DNA viruses with a genome size of approximately 90-160 kb (Murphy et al., 1995). During infection of the host the virus goes through a biphasic cycle producing budded virus (BV) which as its name implies, buds from the cell membrane and an occlusion derived virus (ODV) which becomes incorporated into a crystalline protein matrix, termed an occlusion body (OB), in the nuclei of infected cells. Budded virus is spread within the host by infection of susceptible cells whereas OB are spread by horizontal transmission between susceptible larvae in the environment when the cadavers liquefy and contaminate the surfaces of plants with OB that other susceptible larvae subsequently consume. Infection in the insect is initiated when the protein OB matrix is dissolved in the alkaline midgut lumen releasing virus particles which are taken in by midgut epithelial cells leading to their infection and the spread of the virus within the host.

The widespread use of baculoviruses as biological control agents has been hampered by a number of factors such as host range, persistence in the environment due to UV inactivation, quickness of kill and cost as compared with less expensive chemical insecticides. Another deterrent factor often overlooked in the employment of biological control agents is the uncertainty of not knowing early enough following exposure to the biopesticide whether or not the insects are indeed infected. This is obviously not a problem with the faster acting chemical insecticides.

In the present study we wanted to determine whether the expression of a red fluorescent protein (RFP) gene in susceptible larvae fed OB from a recombinant baculovirus could be used to predict larval infection at an early period following feeding.

Materials and Methods

The virus employed in these studies was a recombinant AcMNPV hsp70 Red CL15 (McIntosh et al. 2005) that was propagated in a *Heliothis subflexa* (BCIRL-Hs-AM1) cell line (McIntosh and Ignoffo, 1989) and the OB harvested and enumerated as previously reported (McIntosh et al., 2006).

First and third instars of *Trichoplusia ni* and *H. subflexa* procured from in-house colonies and *Helicoverpa zea* (Bio-Serv, Frenchtown, NJ) were fed OB concentrations of 1×10^3 /0.1ml and 1×10^5/0.1ml by topical application to a wheat-soy diet (Bio-Serv, Frenchtown, NJ) surface that was allowed to dry before placement of larvae on the diet surfaces. Samples consisted of 50 larvae per OB concentration for each time period with larvae distributed singly on the inoculated surface of each well of a 50 well plastic tray that was covered with mylar, heat sealed, perforated and incubated at 29°C.

At each time period individual larvae were removed to a Petri dish and examined under UV-light for red fluorescence employing an Olympus CK2 inverted microscope with a BH2-RFC attachment and 10 x objective. The number of larvae displaying red fluorescence for each time period was recorded and the percentage of fluorescent larvae calculated. Each larva examined for fluorescence was returned to the tray which was re-sealed and re-incubated. All larvae showing fluorescence were followed for mortality. Controls consisted of larvae fed diet only.

To obtain a comparative measure of observed red fluorescence at various time periods following feeding of recombinant OB by 1st and 3rd instar larvae of the three species, the percent fluorescence was regressed on time in hours following exposure to the two concentrations of OB used in this study. The fluorescent fifty (FL_{50}) was calculated based on the linear regression equation which is the time in hours for 50% of the exposed population to show fluorescence.

Conclusion

Table 1 presents the results of first instar larvae of *T. ni*, *H. subflexa* and *H. zea* fed OB at two concentrations of 1×10^3 and 1×10^5. *T. ni* exposed to the higher concentration of OB showed 100% fluorescence as early as 16h post exposure (pe) to OB and in the second experiment 100% fluorescence occurred at 24h after exposure. *H. subflexa* achieved 100% fluorescence at the higher OB concentration in one experiment at 24h and in the second experiment at 36h pe to OB. At the lower OB concentration 98%-100% fluorescence was observed for *T. ni*, and *H. subflexa* at 24h pe. In contrast *H. zea* which is known to be less susceptible to AcMNPV (Vail et al. 1978; Washburn et al. 1996) gave 100% fluorescence (1st instar) at 48h pe at the higher dose of 1 x 10^5 OB and attained 86%-90% fluorescence at 96h pe at the lower dose of OB. An *H. zea* larva is illustrated in Fig.1c and it shows fluorescence intensity similar to that of the other two species (Fig. 1a,b)

Table 1. Detection of fluorescence in 1st instar larvae fed occlusion bodies (OB) from AcMNPVhsp70-Red recombinant

Insect	Dose (OB/ 0.1 ml)	8 + - %FL	16 + - % FL	24 + - % FL	36 + - % FL	48 + - % FL	72 + - % FL	96 + - % FL	% Mortality
				Time post exposure (h)					
T. ni	1 x 10^3	*1) 8 42 16 *2) 10 40 20	1) 16 34 32 2) 32 18 64	1) 49 1 98 2) 50 0 100	1) 49 1 98	1) 50 0 100			100
T. ni	1 x 10^5	1) 5 45 10 2) 16 34 32	1) 26 24 52 2) 50 0 100	1) 50 0 100					100
H. subflexa	1 x 10^3	1) 2 48 4 2) 0 50 0 0	1) 25 25 50 2) 7 43 14	1) 50 0 100 2) 47 3 94					100
H. subflexa	1 x 10^5	1) 20 30 40 2) 25 25 50	1) 34 16 68 2) 39 11 78	1) 42 8 84 2) 50 0 100	1) 50 0 100				100
H. zea	1 x 10^3	1) NT 2) 0 50 0 0	1) NT 2) 0 50 0 0	1) 5 45 10 2) 4 46 8		1) 10 40 20 2) 15 35 30	1) 14 36 28 2) 32 18 64	1) 43 7 86 2) 45 5 90	90
H. zea	1 x 10^5	1) 6 44 12 2) 6 44 12	1) 22 28 44 2) 24 26 48	1) 38 12 76 2) 36 14 72	1) 45 5 90 2) 32 18 64	1) 50 0 100 2) 50 0 100			100

*1) and 2) represent separate experiments; 50 larvae were analyzed for each time point.
+ = positive fluorescence.
- = no fluorescence.
% FL = percent fluorescence.
NT = not tested.

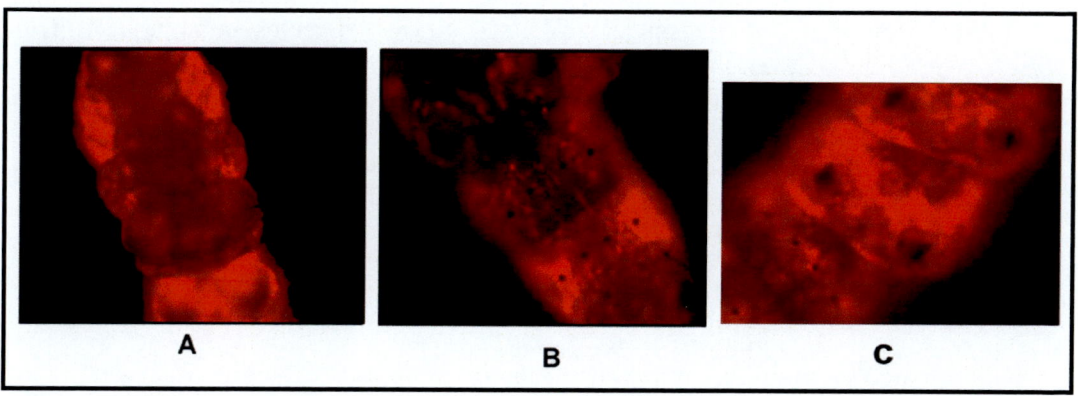

Figure 1. The fluorescence of larvae 24h-36h post-feeding on diet surface inoculated with 1×10^5 OB of the recombinant Achsp70 Red CL15 fed to 24h old larvae. Photomicrographs at 100 x magnification viewed under UV light for red fluorescence with an Olympus CK2 microscope. A) *T. ni* larva B) *H. subflexa* larva and C) *H. zea* larva.

Table 2 illustrates the results of third instar larvae exposed to both doses of OB. *T. ni* fed 1×10^3 OB gave 100% fluorescence between 24h – 48h pe whereas *H. subflexa* attained 100% fluorescence at 72h. At the higher dosage of 1×10^5 OB, 100% fluorescence was obtained for third instar *T. ni* at 24h pe whereas *H. subflexa* attained 100% fluorescence between 36h – 72h pe. Third instar *H. zea* larvae never achieved 100% fluorescence at either OB dose attaining 26-28% fluorescence at 96h pe to the lower dose and 56-66% fluorescence at the higher dose for the same time period.

Fluorescence was first detected in larvae at the first time point tested in this study at 8h following exposure to OB for both first instar *T. ni* and *H. subflexa* fed both concentrations of recombinant OB and ranged from 4% to 50% whereas fluorescence in first instar *H. zea* was not detected until 24h pe (8-10%) for the lower OB dose and 8h (12%) for larvae fed on the higher OB dose (Table 1).

Third instar *T. ni* and *H. subflexa* fed the higher dose of 1×10^5 OB (Table 2) displayed 8-20% fluorescence at 8h pe whereas third instar *H. zea* did not display fluorescence (2-6%) until 36h pe. At the lower OB dosage *T. ni* showed 14-26% fluorescence at 8h pe whereas *H. subflexa* showed 4-6% fluorescence 16h pe. In contrast, *H. zea* third instar showed 2-6% fluorescence at 36h for the higher OB dosage of 1×10^5 and 2-4% fluorescence at 48h for the lower OB dosage of 1×10^3.

In mortality studies, the more susceptible species, namely *T. ni* and *H. subflexa* that showed 100% fluorescence (1st and 3rd instars) also resulted in 100% mortality in those larvae fed both OB concentrations (Tables 1 and 2). The less susceptible *H. zea* produced 90% and 100% mortality for first instar larvae at the low and high OB concentration respectively. In contrast third instar *H. zea* larvae achieved only 54% and 59% mortality at the low and high OB concentration respectively.

Table 2. Detection of fluorescence in 3rd instar larvae fed occlusion bodies (OB) from AcMNPVhsp70-Red recombinant

Insect	Dose(OB/ 0.1 ml)	8 +-% FL	16 +-% FL	24 +-% FL	36 +- % FL	48 +- % FL	72 +-% FL	96 +- % FL	% Mortality
T. ni	1×10^3	*1) 13 37 26 *2) 7 43 14	1) 16 34 32 2) 14 36 28	1) 41 9 82 2) 50 0 100	1) 45 5 90	1) 50 0 100			100
T. ni	1×10^5	1) 10 40 20 2) 7 43 14	1) 17 33 34 2) 15 35 30	1) 50 0 100 2) 50 0 100					100
H. subflexa	1×10^3	1) 0 50 0 2) 0 50 0	1) 3 47 6 2) 2 48 4	1) 12 38 24 2) 11 39 22	1) 28 22 56 2) 29 21 58	1) 38 12 76 2) 39 11 78	1) 50 0 100 2) 50 0 100		100
H. subflexa	1×10^5	1) 6 44 12 2) 4 46 8	1) 15 35 30 2) 12 38 24	1) 32 18 64 2) 28 22 56	1) 50 0 100 2) 44 6 88	2) 48 2 96	2) 50 0 100		100
H. zea	1×10^3	1) - 2) -	1) - 2) -	1) 0 50 0 2) 0 50 0	1) 0 50 0 2) 0 50 0	1) 1 49 2 2) 2 48 4	1) 10 40 20 2) 7 43 14	1) 14 36 28 2) 13 37 26	54
H. zea	1×10^5			1) 0 50 0 2) 0 50 0	1) 3 47 6 2) 1 49 2	1) 10 40 20 2) 8 42 16	1) NT 2) NT	1) 33 17 66 2) 28 22 56	59

*1) and 2) represent separate experiments; 50 larvae were analyzed for each time point.
+ = positive fluorescence.
- = no fluorescence.
% FL = percent fluorescence.
NT = not tested.

Fluorescent fifty (FL_{50}) values, the time taken to attain 50% fluorescence of exposed larvae are depicted in Figures 2 and 3. The shortest time for first instars at the lower OB dose was recorded for *T. ni* (FL_{50} = 15h) followed by *H. subflexa* (FL_{50} = 16.9h) and *H. zea* (FL_{50} = 66.1h). At the higher dose of 1×10^5 OB the FL_{50} values were 12.8h, 28h and 73.6h for *T.ni*, *H. subflexa* and *H. zea* respectively. Third instars exposed to the lower OB dose of 1×10^3 followed the same pattern as the first instars for FL_{50} values with *T. ni* showing the lowest FL_{50} = 18.4h, followed by *H. subflexa* (FL_{50} = 37.4h) and *H. zea* (FL_{50} > 96h). The FL_{50} values at the higher OB dose were 16.0h (*T. ni*), 24h (*H. subflexa*) and 87.1h (*H. zea*). The larger FL_{50} values observed for both first and third instar *H. zea* is probably a reflection of a lower degree of susceptibility of this lepidopteran pest to AcMNPV (Vail et al. 1978; Washburn et al. 1996).

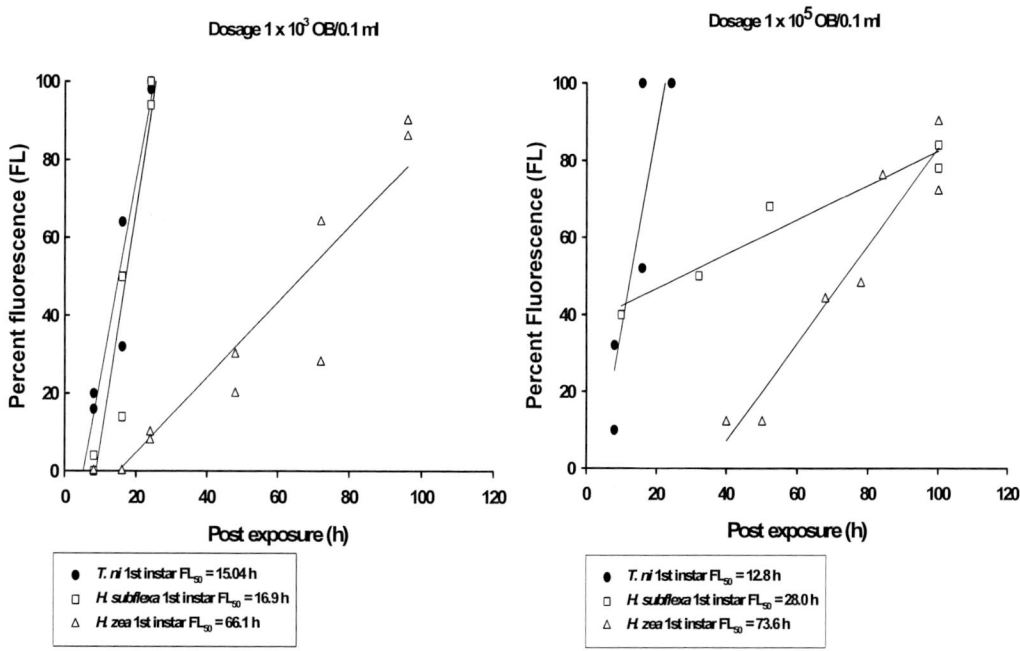

Figure 2. Linear regression and FL_{50} values for *T. ni*, *H. subflexa* and *H. zea* first instar fed 1×10^3 and 1×10^5 OB.

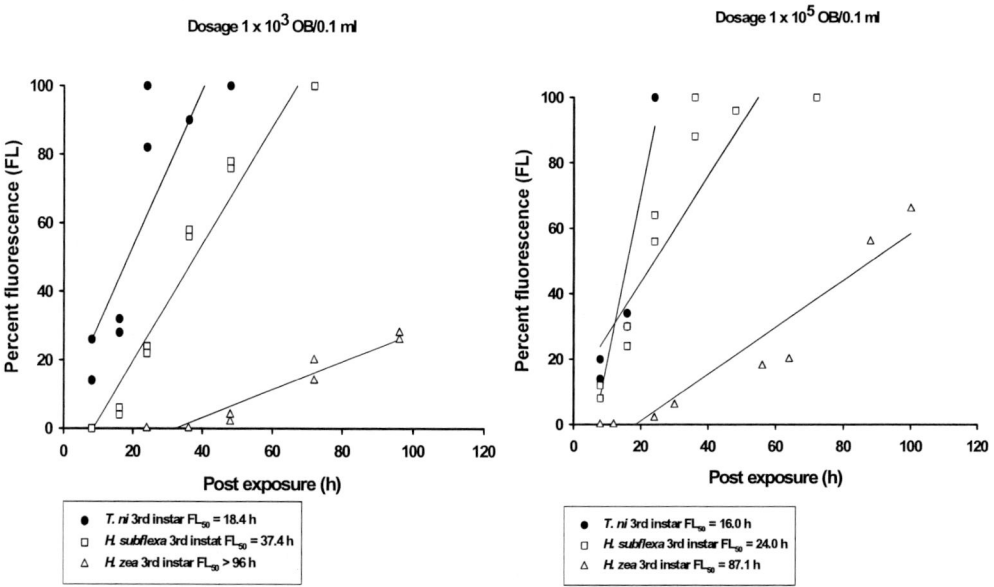

Figure 3. Linear regression and FL_{50} values for *T. ni*, *H. subflexa* and *H. zea* third instar fed 1×10^3 and 1×10^5 OB.

The difference in time taken to attain 100% fluorescence between experiments within a species for first or third instar larvae exposed to the same concentration of OB may be a reflection of a number of factors such as how soon the larvae started to feed after placed on the diet; the extent of feeding which would influence the amount of OB consumed from the diet and thus virus particles that are released and initiate infection; individual variation between larvae as well as physiological conditions such as molting. These factors may also account for some of the FL_{50} values at the higher OB concentration exceeding those at the lower OB concentration for some species.

The present study illustrates that it is possible to detect early expression of the RFP gene and to correlate it with the degree of infection in larvae exposed to a recombinant baculovirus carrying a red fluorescent protein (RFP) gene under the control of an early promoter. The success of this methodology is directly related to the strength and early expression of the hsp70 promoter and the ready detection of red fluorescence under UV. As seen in Fig. 1 expression of RFP was very bright and did not fade appreciable over the course of the experiments making it readily detectable in exposed larvae. Since it is possible to have expression of the gene marker without viral replication (McIntosh et al. 2005) it was necessary to determine whether or not larvae giving 100% fluorescence also resulted in 100% mortality. The results of the present study support this premise.

This methodology could be applied to assess the degree of infection of exposed susceptible larvae under field conditions since the technique is simple only requiring a recombinant carrying a fluorescent marker gene under an appropriate promoter and a suitable UV source to determine fluorescence. What was more remarkable was the detection of fluorescent larvae fed the lower dose of 1×10^3 OB/0.1ml which translates to approximately 1 OB/mm^2 of diet surface. Control larvae (uninfected) did not fluoresce but fras did and was easily distinguished from larvae. Other investigators (Langridge et al. 1996) have employed very sensitive methodologies but more complex and elaborate using baculovirus recombinants carrying the luciferase gene and analysis by low-light video image analysis to follow infection in both insect cell culture and larvae. Bioluminescence was first detected in *T. ni* larvae fed recombinant OB at 72h post feeding (Langridge et al. 1996).

Acknowledgment

We thank Mr. Steve Saathoff for his excellent technical assistance.

References

Boyce, F. M., Bucher, N. L. R. 1996. Baculovirus-mediated gene transfer into mammalian cells. *Proc. Nat. Acad. Sci.* 93, 2348-2352.

Fauquet, C. M., Mayo, M. A., Maniloff, J., Desselberger, U., Ball, L. A. 2005. *Virus Taxonomy*. Academic Press.

Hammock, B. D., Fowler, E. 1991. Insecticidal effects of an insect-specific neurotoxin expressed by a recombinant baculovirus. *Virology* 184, 777-780.

Jehle, J. A., Blissard, G. W., Bonning, B. C., Cory, J. S., Herniou, E. A., Rohrmann, G. F., Theilmann, D. A., Thiem, S. M., Vlak, J. M. 2006. On the classification and nomenclature of baculoviruses: a proposal for revision. *Arch. Virol.* 151, 1257-1266.

Kost, T. A., Condreay, J. P., Jarvis, D. L. 2005. Baculovirus as versatile vectors for protein expression in insect and mammalian cells. *Nature Biotechnology* 23 567-575.

Langridge, W. H. R., Krausova, V. I., Szalay, A. A., Fodor, I. 1996. Detection of baculovirus gene expression in insect cells and larvae by low light video image analysis. *J. Virol. Methods* 61, 151-156.

Luckow, V. L., Summers, M. D., 1988. Trends in the development of baculovirus expression vectors. *Biotechnology* 6, 47-55.

Maeda, S., Volrath, S. L., Hanzlik, T. N., Harper, S. A., Maddox, D. W., Hammock, B. D., Fowler, E. 1991. Insecticidal effects of an insect-specific neurotoxin expressed by a recombinant baculovirus. *Virology* 184, 777-780.

McCutchen, B. F., Choudary, P. V., Crenshaw, R., Maddox, D., Kamita, S. G., Palekar, N. Volrath, S., Fowler, E., Hammock, B. D., Maeda, S. 1991. Development of a recombinant baculovirus expressing an insect selective neurotoxin: potential for pest control. *Biotechnology* 9, 848-852.

McIntosh, A. H., Grasela, J. J., Popham, H. J. R. 2005. AcMNPV in permissive, semipermissive and nonpermissive cell lines from Arthropoda. *In Vitro Cell. Dev. Biol.* 41, 298-304.

McIntosh, A. H., Ignoffo, C. M. 1989. Replication of *Autographa californica* nuclear polyhedrosis virus in five lepidopteran cell lines. *J. Invertebr. Pathol.* 54, 97-102.

McIntosh, A. H., Grasela, J. J., Goodman, C. L. 2006. A simplified and rapid method for extraction of DNA from baculovirus occlusion bodies. *BioProcess J* 5, 59-61.

Murphy, F. A.; Fauquet, C. M.; Bishop, D. H. L.; Ghabrial, S. A.; Jarvis, A. W.; Martelli, G. P.; Mayo, M. A.; Summers, M. D. Virus taxonomy. *Classification and nomenclature of viruses.* Springer-Verlag ; 1995.

O'Reilly, D. R., Miller, L. K., Luckow, V. A., 1992. *Baculovirus Expression Vectors: A Laboratory Manual.* W. H. Freeman, New York.

Tomalski, M. D., Miller, L. K. 1991. Insect paralysis by baculovirus-mediated expression of a mite neurotoxin gene. *Nature* 352, 82-85.

Vail, P. V., Jay, D. L., Stewart, F. D., Martinez, A. J., Dulmage, H. T. 1978. Comparative susceptibility of *Heliothis virescens* and *H. zea* to the nuclear polyhedrosis virus isolated from *Autographa californica*. *J. Econ. Entomol.* 71, 293-296.

Washburn, J. O., Kirkpatrick, B. A., Volkman, L. A. 1996. *Insect protection against viruses.* 383, 767.

Short Communication B

Reovirus-like Double Stranded RNA Fractions in a *Drosophila Melanogaster* Line Containing Individual Second Chromosome from Natural Population

E. G. Pasyukova[1] and D. V. Mukha[2]

[1] Institute of Molecular Genetics of Russian Academy of Sciences, Moscow, Russia;
[2] ~~N. I.~~ Vavilov~~'s~~ Institute of General Genetics of Russian Academy of Sciences, Moscow, Russia

Abstract

DNA samples extracted from 20 *Drosophila melanogaster* lines containing individual second chromosomes from Raleigh population in the genetic background of the isogenic laboratory line Samarkand (Samarkand;Raleigh$_i$;Samarkand) were analyzed using electrophoresis in agarose gel. In two samples, in addition to high molecular weight genomic DNA, ten separate bands of nucleic acid material were found. One of the two lines was analyzed in more detail. All additional bands were resistant to DNAse and RNAse A treatment but sensitive to RNAse III treatment. The size of the bands varied approximately from 1 to 5 kilobases. We hypothesize that these bands represent a double stranded RNA (dsRNA) corresponding to a virus with segmented genome, i.e. a reovirus. ds RNA was present in both males and females of the Samarkand;Raleigh$_{virus}$;Samarkand line and absent in Samarkand flies of both sexes as well as in male and female hybrids of reciprocal crosses between Samarkand and Samarkand;Raleigh$_{virus}$;Samarkand flies. To explain these results, we suggest that replication of cytoplasmic dsRNA fraction transmitted to hybrids by Samarkand;Raleigh$_{virus}$;Samarkand parents is suppressed by the second Samarkand chromosome, while Raleigh$_{virus}$ chromosome may carry a permissive allele of a gene (genes) allowing dsRNA production.. No morphological characters different from wild type were noted in Samarkand;Raleigh$_{virus}$;Samarkand flies. Protocol of dsRNA extraction and purification was proposed. Cloning of putative virus genome for further sequencing and *in situ* hybridization; detection and analysis of virus particles and other experiments aimed to identify and describe the virus are in progress. After virus identification, we consider virus-host genome interactions to be of major interest for further investigation.

Introduction

Only a few viruses of *Drosophila* have been described so far and most of them have not been extensively characterized [Ashburner et al., 2005]. Viruses belonging to different taxa have been found in *Drosophila* cell cultures [Plus, 1978] and in flies: *Drosophila* sigma virus (*Rhabdovitidae*, negative single stranded RNA genome [Fleuriet, 1976; Landes-Devauchelle et al., 1995]); *Drosophila* C virus (DCV, *Dicistroviridae*, positive single stranded RNA genome [Jousset et al., 1977; Johnson, Christian, 1998]); *Drosophila* X virus (DXV, *Birnaviridae*, bisegmented double stranded RNA genome [Teninges et al, 1979; Zambon et al., 2005]); *Drosophila* S virus (DSV, *Reoviridae*, segmented double stranded RNA genome [Louis et al., 1988; Lopez-Ferber et al., 1989]); *Drosophila* Nora virus (*Picornaviridae*-like, positive single stranded RNA genome [Habayeb et al., 2006]); and some others. Recently, explorations into basal mechanisms of innate immune responses, which appeared to be well conserved between humans and insects [Hultmark, 2003; Leclerc, Reichhart, 2004], inspired a renewed interest in *Drosophila* viruses [Zambon et al., 2005; Dostert et al., 2005]. To improve understanding of virus–host genome interactions, better knowledge about various *Drosophila* viruses is needed.

In 2007, for the purpose of analyzing genetic variation of several *Drosophila melanogaster* loci in nature, DNA samples from 20 lines containing individual second chromosomes from Raleigh population in the genetic background of the isogenic laboratory line Samarkand (genotype designated as Samarkand;Raleigh$_i$;Samarkand, [De Luca et al., 2003]) were extracted using standard phenol-chloroform technique [Sambrook et al., 1989] except that RNAse A treatment was added after tissue lysis ("Fermentas", final concentration 100 u/ml, 15 minutes, room temperature). Electrophoresis in 1% agarose gel demonstrated that in two samples, in addition to high molecular weight genomic DNA, at least seven separate bands of nucleic acids material were found. To investigate the nature of this additional material, one of the two lines was analyzed in more detail.

The Nature of the Additional Nucleic Acids Material

First of all it was necessary to reproduce the result, to make sure that additional nucleic acids material was not present in the sample due to some kind of contamination. It was also reasonable to test whether additional bands were present both in male and female flies. To standardize experimental conditions, DNA in this and all further experiments was extracted from 30 15-day old flies. Figure 1 (lane 1 and 2) illustrates that additional bands of nucleic acids material were present both in males and females. Three independent extractions showed the same results indicating that additional bands of nucleic acids material are characteristic of this particular line.

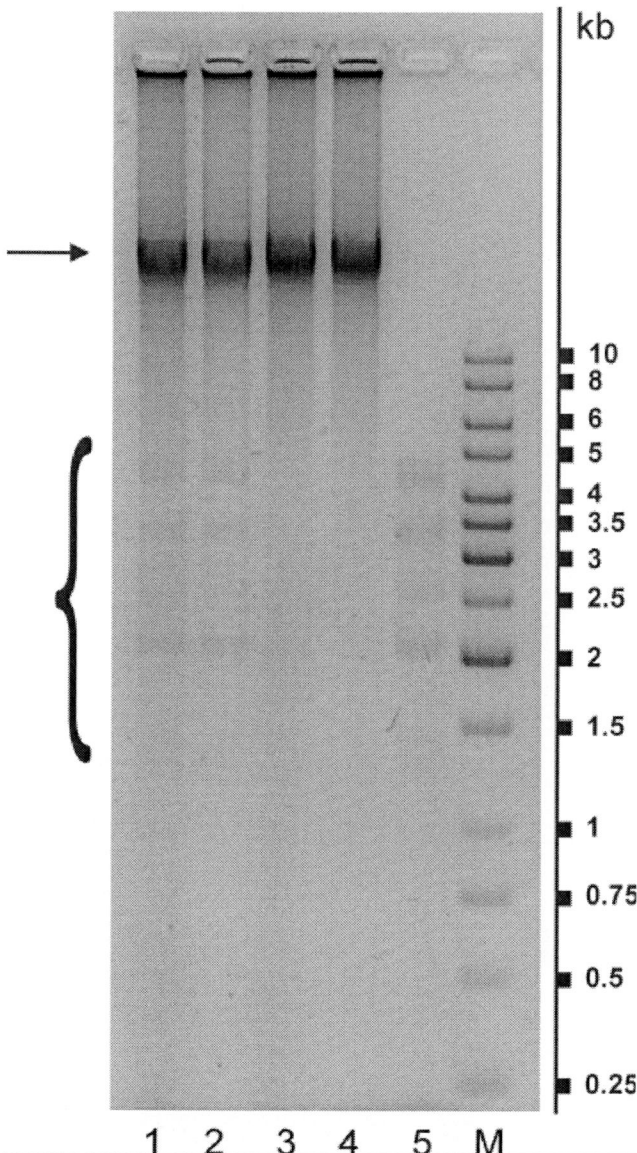

Figure 1. Electrophoretic separation in 1% agarose gel of nucleic acids extracted from flies of different sexes and genotypes.
1 and 2 – females and males of the line containing individual second chromosomes from Raleigh population; 3 and 4 – hybrids (samples from females and males combined) from reciprocal crosses between the line containing individual second chromosomes from Raleigh population and isogenic laboratory line Samarkand; 5 – females plus males of the line containing individual second chromosomes from Raleigh population, after DNAse treatment; M – marker, 1 kb ladder ("Fermentas").
Arrow indicates high molecular weight genomic DNA; crochet indicates bands corresponding to additional nucleic acids material.

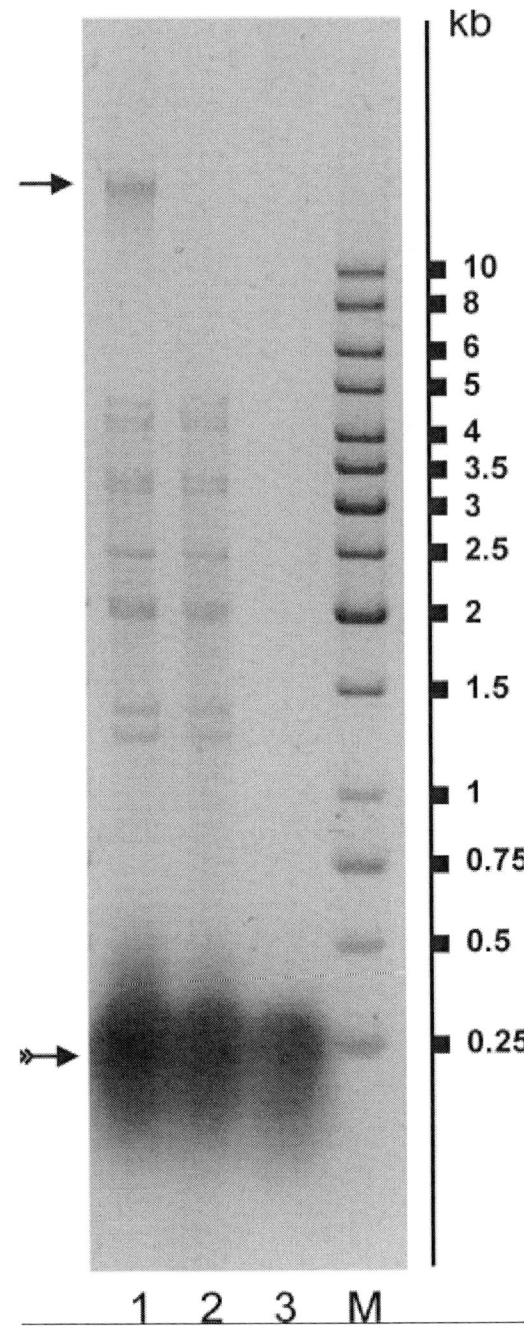

Figure 2. Electrophoretic separation in 1% agarose gel of the diffused fraction of nucleic acids collected after fractionation by equilibrium centrifugation in CsCl-ethidium bromide density gradients (see text for explanations).
1 – initial material; 2 – initial material after DNAse treatment; 3 – initial material after DNAse and RNAse III treatment; M –marker, 1 kb ladder ("Fermentas").
Arrow indicates DNA; feathered arrow indicates mixture of small fragments of nucleic acids.

The protocol of DNA extraction included RNAse A treatment, thus additional nucleic acids material is not single stranded RNA (ssRNA). The mixture of male and female nucleic acids samples (approximately 10 µg) was further treated by DNAse ("Fermentas", final concentration 20 u/ml , 60 minutes, 37°C). Figure 1 (lane 5) demonstrates that high molecular weight DNA was sensitive while additional nucleic acids material was resistant to DNAse treatment, which proves that this material is not DNA. This suggests that additional nucleic acids material is double stranded RNA (dsRNA).

For further analysis, an attempt to purify and enrich additional nucleic acids material was made using fractionation by equilibrium centrifugation in CsCl-ethidium bromide density gradients. Total nucleic acids for centrifugation were extracted from 200 flies using standard phenol-chloroform technique [Sambrook et al., 1989]. Approximately 100 µg of total nucleic acids were dissolved in CsCl-ethidium bromide solution according to the Purification of Closed Circular DNA protocol [Sambrook et al., 1989] and centrifuged at 45,000 rpm for 48 hours (Beckman Spinco L2-65B, Ti50 rotor). After centrifugation, two regular fractions corresponding to linear DNA (in the middle of the tube) and ssRNA (pellet on the bottom of the tube) were revealed under UV light (365 nanometers). In addition, a diffused fraction was observed between the other two. This fraction was collected from the tube, CsCl and ethidium bromide were removed using standard protocols [Sambrook et al., 1989], and nucleic acids were ethanol precipitated and dissolved in water.

The result of electrophoretic separation in 1% agarose gel of dissolved material presumably comprising a mixture of circular DNA and dsRNA is shown on figure 2, lane 1. On the bottom of the gel, a huge band corresponding to small fragments of nucleic acids of unknown nature could be observed. This material may be of interest for further analysis which, however, is beyond the scope of this paper. Several bands in the upper part of the gel had the same pattern as bands corresponding to additional nucleic acids material revealed earlier, and attention was focused on an attempt to further investigate the nature of this material.

To distinguish between DNA and dsRNA, the material was successively treated by DNAse and RNAse III ("Ambion", 10 u/ml , 60 minutes, 37°C). One band with the highest molecular weight was sensitive to DNAse (figure 2, lane 2) and probably corresponds to mitochondrial DNA. All other additional bands except the lowest were resistant to DNAse but sensitive to RNAse III treatment (figure 2, lanes 2 and 3). Consequently, these bands represent dsRNA presumably corresponding to a virus with a segmented dsRNA genome, i.e. a reovirus. For convenience, the line containing fragmented dsRNA will be further referred to as Samarkand;Raleigh$_{virus}$;Samarkand. Important technical conclusion from this experiment is that fractionation in Cesium density gradient can be considered as a new approach to dsRNA extraction from a nucleic acid mixture, which was applied here for the first time.

Higher concentration of additional nucleic acids material after purification and better resolution of electrophoretic separation allowed to discriminate 10 fragments of dsRNA varying in size approximately (considering DNA markers used) from 1 to 5 kilobases. Genomes of reoviruses typically consist of 10-12 dsRNA fragments. DSV genome consists of at least eight, probably nine fragments [López-Ferber et al., 1989]. Reoviruses were also found in *Drosophila* cell cultures [Alatortsev et al., 1981; Hsu, Sanders, 1983]. In the first case 10 dsRNA fragments and in the later case 11-13 dsRNA fragments varying in size from 980 to 4600 base pairs and non-homologous to each other and genomic DNA were found. It is worth noting that it was the presence of fragmented dsRNA that lead authors to the conclusion that viruses belonging to *Reoviridae* family were found in *Drosophila* lines and cell cultures. The number and size of dsRNA fragments described here is different from those described earlier, besides, DSV is characteristic to *Drosophila simulans*, and these facts point toward the possibility that different species of reoviruses could infect *Drosophila*.

It should be stressed, however, that so far it is impossible to assert that fragmented dsRNA found in Samarkand;Raleigh$_{virus}$;Samarkand line belongs to a reovirus. To prove it, cloning of putative virus genome for further sequencing , analysis of virus particles and other experiments aimed to identify and describe the virus are now in progress. Recently, some *Drosophila* viruses of other families were cloned and sequenced (DXV [Chung et al., 1996]; DCV [Johnson, Christian, 1998]; Nora virus [Habayeb et al., 2006]). Unfortunately, common databases do not provide neither references to publications concerning *Drosophila* reoviruses genome sequences, nor any other information on this subject. However, genome sequences of reoviruses of other insects are available and could be used for identification purposes.

It is well known that *Drosophila* viruses may affect important characteristics of flies, such as developmental rate, stress resistance, bristles morphology [Fleuriet, 1996; Louis et al., 1987]. In other cases, obvious effects were not revealed [Habayeb et al., 2006]. No morphological characters different from wild type were noted in Samarkand; Raleigh$_{virus}$; Samarkand flies. Special experiments are needed to find out whether other traits are affected.

Putative Virus-host Genome Interactions

The Samarkand;Raleigh$_{virus}$;Samarkand line differs from the isogenic Samarkand line by the presence of the second Raleigh$_{virus}$ chromosome. Nucleic acids were extracted from Samarkand males and females using the above mentioned standard phenol-chloroform technique and RNAse A treatment, and dsRNA bands were not revealed after separation in agarose gel (data not shown). Furthermore, dsRNA bands were not revealed in male and female hybrids of reciprocal crosses between Samarkand and Samarkand; Raleigh$_{virus}$; Samarkand flies (figure 1, lanes 3 and 4). To explain these results, we suggest that replication of cytoplasmic dsRNA fraction transmitted to hybrids by Samarkand;Raleigh$_{virus}$;Samarkand parents could be suppressed by the second Samarkand chromosome, while Raleigh$_{virus}$ chromosome may carry a permissive allele of a gene (genes) allowing dsRNA production. Recent publications have awakened fresh interest to mechanisms of antiviral control effected by host genome. It was shown that the Toll pathway is involved in inhibition of DXV replication, while JAK/STAT pathway plays a role in response against DCV [Zambon et al., 2005; Dostert et al., 2005]. It was also demonstrated that genes playing a key role in RNA interference are involved in antivirus response in *Drosophila* [Wang et al, 2006]. A more complicated explanation of the results presented here needs additional assumption that putative virus is not transmitted by cytoplasm. Sequences corresponding to its genome could be localized in the second Raleigh$_{virus}$ chromosome, and cytoplasmic dsRNA fraction could be synthesized *de novo* in each generation but only in flies homozygous for Raleigh$_{virus}$ because the second Samarkand chromosome suppresses this synthesis. However, up to now, chromosomal localization has only been described for *Drosophila* retroviruses.

From data given above it follows that additional nucleic acids bands were found in only two lines of 20 containing individual second chromosomes from Raleigh population. This fact suggests that in Raleigh population a second chromosome locus controlling putative virus replication is polymorphic for permissive and repressive alleles. *Drosophila* sigma virus was thoroughly studied in natural populations where the frequency of infected flies varied approximately from 4 to 40 percent depending on particular population and the year of analysis [Fleuriet, 1996]. Two alleles, permissive and restrictive, of the *Drosophila* gene were found [Contamine et al., 1989], and polymorphism for these alleles appeared to be a constant characteristic of populations [Fleuriet, 1996].

After virus identification, we consider virus-host genome interactions to be of major interest for further investigation.

Conclusion

Additional nucleic acids material in a *Drosophila melanogaster* line containing an individual second chromosome from Raleigh population represent fragmented dsRNA which most likely corresponds to a virus with fragmented dsRNA genome, i.e. a reovirus.

Fractionation by equilibrium centrifugation in CsCl-ethidium bromide density gradients is proposed as a method allowing purification of dsRNA.

The presence of dsRNA fractions in a nucleic acids sample depends on the genotype of the *Drosophila* line and is associated with the presence of the second chromosome from Raleigh population. This chromosome may carry a permissive allele of a gene (genes) allowing dsRNA production.

Acknowledgments

We are grateful for Trudy Mackay for providing *Drosophila* lines for analysis. This work was supported by RFBR grant # 06-04-49453-a and RAS Programe "Biodiversity" for EGP and RFBR grant # 08-04-01402-a and RAS Programe "Biodiversity" for DVM.

References

Alatortsev VE; Ananiev EV; Guschina EA; Grigoriev VG; Guschin BV. A virus of the *Reoviridae* in established cell lines of *Drosophila melanogaster*. *J. Gen. Virol.*, 1981 54:23-31.

Ashburner M; Golik KG; Hawley RS. *Drosophila*: a Laboratory Handbook. Second edition. NY: Cold Spring Harbor Laboratory Press; 2005.

Chung HK; Kordyban S; Cameron L; Dobos P. Sequence analysis of the bicistronic *Drosophila* X virus genome segment A and its encoded polypeptides. *Virology*, 1996 225:359-368.

Contamine D; Petitjean AM; Ashburner M. Genetic resistance to virus infection:the molecular cloning of a *Drosophila* gene that restricts infection by the rhabdovirus sigma. *Genetics*, 1989 123:525-533.

Dostert C; Jouanguy E; Irving P; Troxler L; Galiana-Arnoux D; Hetru C; Hoffman JA; Imler J-L. The JAK-STAT signaling pathway is required but not sufficient for the antiviral response in *Drosophila*. *Nat. Immunol.*, 2005 6:946-953.

Fleuriet A. Presence of hereditary Rhabdovirus sigma and polymorphism for a gene for resistance for this virus in natural populations of *D. melanogaster*. *Evolution*, 1976 30:735-739.

Fleuriet A. Polymorphism of *Drosophila melanogaster* – sigma virus system. *J. Evol. Biol.*, 1996 9:471-484.

Habayeb MS; Ekengren SK; Hultmark D. Nora virus, a persistent virus in *Drosophila*, defines a new picorna-like virus family. *J. Gen. Virol.*, 2006 87:3045-3051.

Hsu JT; Sanders MM. Characterization of a segmented double-stranded RNA virus in *Drosophila* K_c cells. *Nucl. Acids Res.*, 1983 11:3665-3678.

Hultmark D. *Drosophila* immunity: pahs and patterns. *Curr. Opin. Immunol.*, 2003 15:12-19.

Johnson KN; Christian PD. A novel genome organization of the insect picorna-like virus *Drosophila* C virus suggests this virus belongs to a previously undescribed virus family. *J. Gen. Virol.*, 1998 79:191-203.

Jousset F; Bergoin M; Revert B. Characterization of the *Drosophila* C virus. *J. Gen. Virol.*, 1977 34:269-285.

Leclerc V; Reichhart J-M. The immune response of *Drosophila melanogaster*. *Immunol. Rev.*, 2004 198:59-71.

López-Ferber M; Veyrunes JC; Croizier L. *Drosophila* S virus is a member of the *Reoviridae* family. *J. Virol.*, 1989 63:1007-1009.

Louis C; Lopez-Ferber M; Comendador M; Plus N; Kuhl G; Baker S. *Drosophila* S virus, a hereditary reolike virus, probable agent of the morphological S character in Drosophila simulans. *J. Virol.*, 1988 62:1266-1270.

Landes-Devauchelle C; Bras F; Dezelee S; Teninges S. Gene 2 of the sigma rhabdovirus genome encodes the P protein, and gene 3 encodes a protein related to the reverse transcriptase of retroelements. *Virology*, 1995 213:300-312.

Plus N. Endogenous viruses of *Drosophila melanogaster* cell lines: their frequency, identification and origin. *In Vitro*, 1978 14:1015-1021.

Sambrook J; Fritsch EF; Maniatis T. Molecular cloning: a laboratory manual. Second edition. NY: Cold Spring Harbor Laboratory Press; 1989.

Teninges D; Ohanessian A; Richard-Molard C; Contamine D. Isolation and biological properties of *Drosophila* X virus. *J. Gen. Virol.*, 1979 42:241-254.

Wang X-H; Aliyari R; Li W-X; Li H-W; Kim K; Carthew R; Atkinson P; Ding S-W. RNA interference directs innate immunity against viruses in adult *Drosophila*. *Science*, 2006 312:452-454.

Zambon RA.; Nandakumar M., Vakharia VN; Wu LP. The Toll pathway is important for an antiviral response in *Drosophila*. *Proc. Nat. Acad. Sci. USA*, 2005 102:7257-7262.

In: Insect Viruses: Detection, Characterization and Roles
Editors: Ch. J. Connell and D. P. Ralston

ISBN: 978-1-60692-965-0
© 2009 Nova Science Publishers, Inc.

Short Communication C

Cellular Secretion of Sf21 Cells upon Baculovirus Infection

Xiao-Wen Cheng

Department of Microbiology, 32 Pearson Hall, Miami University,
Oxford, Ohio 45056 USA

Baculoviruses have long been used as natural agents to control insect pests in agriculture and forestry due to their notable safety for the environment and public health. Many baculoviruses, specifically the nucleopolyhedroviruses (NPV), have a very narrow host-range, killing one or two insect species. For example, the *Spodoptera exigua* multicapsid NPV (SeMNPV) and *Thysanoplusia orichalcea* multicapsid NPV (ThorMNPV) can kill only one or a few hosts and replicate well in a few insect cell lines (Cheng et al., 2005; Simon et al., 2004; Wang et al., 2008). A noteworthy exception to this generalization is the *Autographa californica* multicapsid NPV (AcMNPV) that has a wide host-range (>33 species) and can replicate well in many insect cell lines (Groner, 1986; McIntosh and Grasela, 1994). AcMNPV is the type species of the family *Baculoviridae*, and is also the first baculovirus whose genome was sequenced (Ayres et al., 1994). AcMNPV has a double-stranded circular DNA genome with a size of 134 kb (Ayres *et al.*, 1994). ThorMNPV and AcMNPV belong to the same group (I) and share about 80% DNA sequence homology (Cheng et al., 2005). SeMNPV belongs to the group II that shares less genome sequence homology to the group I viruses.

A recombinant vThGFP was generated by inserting a green fluorescent protein (GFP) gene expression cassette with an AcMNPV late polyhedrin promoter at the *gp37* locus of ThorMNPV for cell infection studies (Wang et al., 2008). vThGFP infects High 5 (Hi5) derived from cabbage looper *Trichoplusia ni* very well (Cheng et al., 2005; Wang et al., 2008) but replicates poorly in Sf21 cells derived from the fall armyworm *Spodoptera frugiperda*. vThGFP is able to enter Sf21 cells but the viral DNA replication is slow, such that the Sf21 cells containing vThGFP continue cell division and eventually suppress infection by vThGFP. Alternatively, AcMNPV replicates very well in Sf21 cells. When Sf21 cells were co-infected with AcMNPV and vThGFP, AcMNPV helped vThGFP infection rates in Sf21 by 25-fold (Wang et al., 2008). We initially suspected a recombination event between AcMNPV and vThGFP occurred to explain the increased numbers of Sf21 cells infected by vThGFP in the co-infection experiment. Subsequent viral plaque assays using budded virus (BV) from the co-infection and BVs from Hi5 cells infected with vThGFP showed that vThGFP BVs generated from Hi5 cells could not form viral plaques but individual Sf21 cells were infected by vThGFP by showing GFP expression. vThGFP formed plaques in Sf21 cells when BVs from co-infection were used in the plaque assay. Purified vThGFP plaques from the co-infection showed no difference in infecting Sf21 cell compared to the parent vThGFP, and both showed slow infection in Sf21 cells. This suggested that the increase of infection of vThGFP in Sf21 in the co-infection is not due to a recombination event (Wang et al., 2008). Furthermore, use of an AcMNPV cosmid library to co-transfect Sf21 with vThGFP did not reveal any genes from AcMNPV that can enhance vThGFP infection in Sf21 cells (Wang et al., 2008).

To address the question how AcMNPV helps vThGFP replication in Sf21 cells, another hypothesis was proposed; that was AcMNPV infection in Sf21 was able to stimulate secretion from Sf21 cells. Sf21 cells were infected by AcMNPV, and at day 4 post infection, the media were centrifuged to remove BVs in the media. When the centrifuged media were used to incubate Sf21 cells infected with vThGFP, vThGFP infection in Sf21 increased substantially compared to that without the centrifuged media (Wang et al., 2008). This suggests that the centrifuged media may contain secreted materials from Sf21 cells infected by the wide host range AcMNPV that enhanced vThGFP replication in Sf21 cells. Secretion of insect cells infected by baculovirus has never been fully analyzed before, this needs to be investigated in order to understand how virus interacts with cells and changes the permissibility of the cells for viral infection.

All cells secrete in order to communicate with other cells and to maintain the well-being of the cells as well response to environmental changes. In general, eukaryotic cells have two types of secretion: constitutive secretion or house keeping secretion and regulated secretion. Constitutive secretion is required by the cells to maintain their well-being but regulated secretion is needed for the cells to adjust the metabolism of the cells to environmental challenges. When viruses infect mammalian cells, the cells secrete interferon so that other cells become immune to viral infection. However, viruses are not left helpless because the cells mount a defense by secretion of anti-viral proteins. Otherwise viruses should have been eliminated in the environment. Little has been addressed on insect cellular secretion. Nevertheless, it was reported that a homologue of cathepsin L is secreted by non infected Sf9 cells in the conditioned media (Lindskog, Svensson, and Haggstrom, 2006). A metalloproteinase was found in the day 2 and 3 conditioned serum free media of *Trichoplusia ni* (BTI-Tn-5B1-4, High 5) cells and this metalloproteinase is able to shorten the lag phase and increase the maximum cell growth density of *T. ni* cells (Eriksson et al., 2005). One area that has never been fully addressed in the past is how baculovirus infection in insect cells changes the secretome of insect cells.

To understand the insect cellular secretion upon baculovirus infection, we performed a preliminary experiment by infecting Sf21 cells with AcMNPV. In order to compare secretion profiles of Sf21 cells infected by AcMNPV, the Sf21 cells were grown in HyQ SFX-Insect serum free media (HyClone). This eliminates potential fetal bovine serum in the traditional cell culture media that interfere with data analysis. Sf21 cells were infected with AcMNPV at a multiplicity of infection (MOI) of 10 viruses/cell and infected cells were incubated in HyQ SFX at 27 °C for 48 h (O'Reilly, Miller, and Luckow, 1992). Control Sf21 cells were mock infected with HyQ SFX media and incubated in HyQ SF media. Media from virally infected (CMV) and mock infected Sf21 cells (CM) were harvested and centrifuged at 196,408 x g for 15 hr at 4 °C to remove BVs in the CMV. The centrifuged media were lyophilized and analyzed by two-dimensional (2-D) gel electrophoresis and the gels were silver stained to visualize secreted soluble proteins from Sf21 cells infected by AcMNPV.

Silver staining of 2-D gel electrophoresis revealed that when Sf21 cells were infected by AcMNPV, the cells stopped secretion of certain proteins into the media but additional proteins were secreted following AcMNPV infection at 48 h post infection (Figure 1 A and B). It is also noted that majority of the secreted proteins maintained after Sf21 cells were infected by AcMNPV at 48 h post infection. At this time, we still don't know if these secreted proteins of Sf21 cells following AcMNPV infection are the responsible agents for the enhanced viral replication of vThGFP in Sf21 cells. We don't know if the additional proteins in the Sf21 cell media after AcMNPV infection are of cellular or viral origins either. These additional proteins secreted by Sf21 cells following AcMNPV infection need to be identified by peptide sequencing or mass spectrometry in order to fully understand what Sf21 cells secrete upon a viral infection.

Acknowledgments

I would like to thank Drs. T. Z. Salem and L.-H. Wang for providing experimental data for this article during their work in my lab. Dr. D. E. Lynn is credited for critically reviewing this manuscript. Thanks are also due to Drs. K. Z. Abshire and S.-J. Guo for help with 2-D analysis.

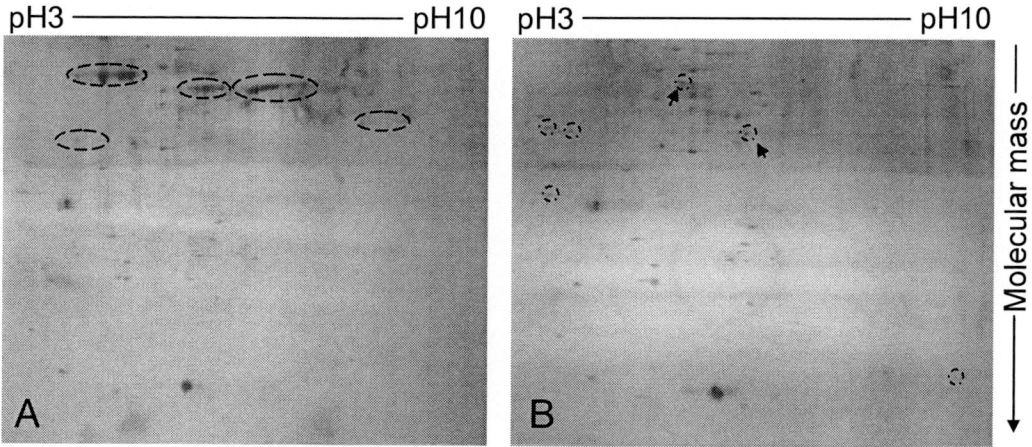

Figure 1. Secretome profile of Sf21 cells infected by AcMNPV. A) Secretome of Sf21 without viral infection at 48 h post infection. ⟃⟆ designates to secreted proteins disappeared in media after Sf-21 cells were

infected by AcMNPV. B) secreome of Sf21 cells infected by AcMNPV at 48 h post infection. designates to additional proteins in the media after Sf21 cells were infected by AcMNPV.

References

Ayres, M. D., Howard, S. C., Kuzio, J., Lopez-Ferber, M., and Possee, R. D. (1994). The complete DNA sequence of *Autographa californica* nuclear polyhedrosis virus. *Virology* 202(2), 586-605.

Cheng, X. W., Carner, G. R., Lange, M., Jehle, J. A., and Arif, B. M. (2005). Biological and molecular characterization of a multicapsid nucleopolyhedrovirus from *Thysanoplusia orichalcea* (L.) (Lepidoptera: Noctuidae). *J Invertebr Pathol* 88(2), 126-35.

Eriksson, U., Hassel, J., Lullau, E., and Haggstrom, L. (2005). Metalloproteinase activity is the sole factor responsible for the growth-promoting effect of conditioned medium in *Trichoplusia ni* insect cell cultures. *J Biotechnol* 119(1), 76-86.

Groner, A. (1986). Specificity and safety of baculoviruses. *In* "The biology of baculovirueses" (R. R. Granados, and B. A. Federici, Eds.), Vol. 1, pp. 177-202. CRC Press, Boca Raton.

Lindskog, E., Svensson, I., and Haggstrom, L. (2006). A homologue of cathepsin L identified in conditioned medium from Sf9 insect cells. *Appl Microbiol Biotechnol* 71(4), 444-9.

McIntosh, A. H., and Grasela, J. J. (1994). Specificity of baculoviruses. *In* "Insect Cell Biotechnology" (M. A. H. Maramorosch K, Ed.), pp. 57-69. CRC Press, Boca Raton.

O'Reilly, D. R., Miller, L. K., and Luckow, V. A. (1992). "Baculovirus expression vectors: a laboratory manual." W. H. Freeman & Co., New York.

Simon, O., Williams, T., Lopez-Ferber, M., and Caballero, P. (2004). Virus entry or the primary infection cycle are not the principal determinants of host specificity of *Spodoptera* spp. nucleopolyhedroviruses. *J Gen Virol* 85(Pt 10), 2845-55.

Wang, L., Salem, T. Z., Lynn, D. E., and Cheng, X. W. (2008). Slow cell infection, inefficient primary infection and inability to replicate in the fat body determine the host range of *Thysanoplusia orichalcea* nucleopolyhedrovirus. *J Gen Virol* 89(Pt 6), 1402-10.

Index

A

abdomen, 42, 51
abdominal, 105
abiotic, 5, 10, 17
abnormalities, 90, 106, 109
accidents, 95
accounting, ix, 103, 106
acetate, 124
acid, xi, 4, 11, 18, 26, 45, 113, 161, 165
acquired immunity, ix, 78, 79
actin, 30, 37
action potential, 110, 111
activation, 25, 26, 27, 29, 30, 35, 51, 62
acute, viii, 45, 49, 53, 54, 56, 59, 60, 71, 72, 73, 74, 100, 106, 108, 109, 110, 111, 112, 114, 115, 138
acute febrile illness, 108, 112
acute infection, 114, 115
Adams, 5, 13, 141
adaptability, 56
adaptation, ix, 78, 79
additives, 2, 12
adenovirus, 112
adhesion, 30, 31, 33
administration, vii, 1, 3, 5, 11
adult, viii, 29, 33, 41, 49, 52, 60, 72, 84, 123, 124, 125, 126, 142, 145, 169
adult population, 84
adults, 42, 49, 88, 90, 92, 96, 98, 125, 129
Aedes, 22, 27, 33, 34, 81, 84, 86, 92, 96, 100
aerosol, 91
aetiology, 94
affect, x, 121, 122, 123, 126
Africa, ix, 26, 78, 79, 84, 85, 87, 97, 99, 104
AGC, 124

age, 42, 55, 90, 107, 108, 109, 113, 114, 116, 151
agent, x, 78, 81, 82, 84, 89, 116, 121, 122, 123, 129, 131, 135, 136, 138, 139, 140, 141, 142, 143, 144, 146, 147, 168
agents, ix, x, xi, 10, 14, 26, 78, 79, 81, 88, 89, 100, 115, 120, 121, 122, 123, 124, 125, 128, 129, 134, 135, 136, 137, 138, 140, 145, 146, 152, 171, 173
aggression, 42, 43, 50
aggressive behavior, 47
aggressiveness, 57
aging, 85
agricultural, 43, 60, 61, 74, 79
agricultural crop, 43
agriculture, vii, xi, 1, 2, 10, 60, 61, 171
aid, 109
Alabama, 98
Alaska, 79, 102
alcohol, 5
Algeria, 84
alkaline, 6, 152
allele, xi, 113, 161, 166, 167
alleles, 167
alpha, 116, 120
alternative, 12, 13, 29, 32, 71
alternatives, 23
alters, 18, 79, 146
altruistic behavior, 42, 43
amino, 4, 26, 45
amino acid, 4, 26, 45
aminopeptidase, 37
amniotic, 94
amphibians, 80, 81
anatomy, 119
anemia, 105
animal models, 99

animals, vii, viii, ix, x, 2, 21, 23, 30, 34, 35, 41, 42, 49, 56, 80, 90, 93, 103, 104, 108, 121, 122, 124, 128, 135, 136, 137, 140, 141
annual, 85
Anopheles gambiae, 22, 33, 35, 36
ANOVA, 93, 94
antagonistic, 122, 137
antibacterial, 25, 35
anti-bacterial, 29
antibiotic, 14
antibodies, 92, 102, 144
antibody, 93, 96, 106, 114, 137
antibody (Ab), 93, 96
antigen, 131, 144
antineoplastic, 120
antineoplastic agents, 120
antiviral, 24, 32, 34, 35, 36, 55, 113, 115, 116, 166, 168, 169
antiviral agents, 115
antiviral therapy, 115
aphasia, x, 103, 112, 113, 118, 119
apoptosis, 26, 35, 36, 91
application, viii, 21, 57, 62, 145, 153
applied research, vii, 1, 2
Arabia, 98
arbovirus infection, 97, 98, 100
Arboviruses, v, viii, 77, 78, 80, 88, 96, 99
arthralgia, 105
arthritis, 135, 146
Arthropoda, 159
arthropods, vii, 10, 38, 78, 79, 91
aseptic, x, 80, 81, 90, 103, 108, 109, 110, 118
aseptic meningitis, x, 80, 81, 90, 103, 108, 109, 110, 118
Asia, ix, 47, 69, 78, 79, 83, 85, 86, 88
Asian, 46, 102
aspiration, 145
assessment, 73, 142
asymptomatic, viii, 65, 77, 78, 79, 89
ataxia, 90, 111, 112
ATC, 124
Atlantic, 82, 83
atrophy, 106
attachment, 104, 113, 114, 145, 153
attacker, 43
attacks, 90
Australia, 88, 89, 95, 97, 101
Austria, 85, 87, 94
autoimmune, 111
autonomic, 114

autopsy, 107, 110, 113
availability, 22
axon, 111
axonal, 109, 110
axons, 114

B

babesiosis, 144
babies, 104
Bacillus, 37
Bacillus thuringiensis, 37
backcross, 57
bacteria, 6, 24, 25, 26, 27, 28, 29, 30, 31, 34, 36, 43
bacterial, 6, 12, 25, 26, 27, 29, 82, 135, 136, 142, 143
bacterial infection, 27, 29, 142
bacteriostatic, 25
bacterium, 19, 28, 32
barrier, vii, 1, 2, 6, 24, 27, 50, 91, 113
barriers, 24
basal ganglia, 90
basal nuclei, 91
base pair, 166
battery, 25
BBB, 113, 114
beetles, 24, 27
behavior, 42, 43, 47, 49, 51, 54, 55, 56, 58, 137
behavioral change, 49, 52
behavioral effects, 50
Belgium, 142
benign, 81, 82, 88
bilateral, 110, 111, 112
binding, 6, 9, 10, 15, 17, 18, 19, 25, 26, 30, 33, 35
bioassays, 4, 8, 10
biological activity, 15
biological control, 10, 152
biological control agents, 10, 152
biological processes, 30
biology, ix, 78
biopsy, 146
birds, ix, 80, 81, 84, 86, 88, 103, 104, 108, 116, 117, 118, 120
bladder, 109
Blattella germanica, 22, 36
bleeding, 27, 35
blocks, 27, 31
blood, 24, 25, 27, 34, 91, 104, 113, 116, 117, 131, 132, 133, 135, 136, 146, 147
blood clot, 24

blurring, 111
body size, 55, 60
body temperature, 93, 94
bootstrap, 46
borderline, 92, 93
boric acid, 18
Borrelia infection, 125, 131, 134, 135, 142
borreliosis, 92, 96, 140, 142, 144, 146
bovine, 146, 173
brain, 51, 90, 91, 99, 105, 107, 112, 113, 114, 115
brain stem, 90, 91, 112, 114
brainstem, 90, 119
Brazil, 21, 81, 96
Brazilian, 69, 75, 96
breastfeeding, 104
broad spectrum, 25
buffer, 62, 124
Bulgaria, 140

C

cabbage, 15, 171
Caenorhabditis elegans, 23, 34, 38
calcium, 38
California, 141, 142, 143
campylobacter, 111
Canada, 60, 81, 82, 83, 95
Cancer, 120
Candida, 30, 37
capillaries, 91
capsule, 15, 19, 25
cardiopulmonary, 112
Caribbean, 98
Caribbean Islands, 98
carrier, 34
caterpillars, 28
cattle, 98, 122, 136, 139
causality, 50
CD4, 107
cDNA, 22, 26, 28, 30, 45, 46, 50, 62
celery, 15
cell, viii, xi, 4, 9, 16, 24, 26, 30, 37, 46, 50, 53, 54, 55, 56, 59, 60, 73, 74, 91, 94, 113, 114, 115, 131, 144, 152, 153, 157, 159, 162, 165, 168, 171, 172, 173, 174
cell culture, 4, 131, 144, 157, 162, 165, 173, 174
cell death, 91
cell division, 171
cell growth, 172
cell line, xi, 50, 56, 94, 153, 159, 168, 171

cell lines, xi, 56, 94, 159, 168, 171
cell surface, 30
cells, 90, 91, 93, 94
Centers for Disease Control, 119, 121
Central African Republic, 85
Central America, 81, 83
Central Europe, 85, 86, 87, 143
central nervous system (CNS) ix, 81, 84, 86, 89, 91, 96, 99, 100 , 103, 105, 106, 107, 111, 113, 114, 115, 116, 118
cerebellar ataxia, 111
cerebellum, 91, 107
cerebral cortex, 91
cerebrospinal fluid (CSF), 91, 106, 108, 109, 110, 111, 112, 113, 114
cerebrum, 107, 114
Cetyltrimethylammonium bromide, 11
chemotherapy, 107, 112, 114, 115
Chicago, 120
chicken, 94
children, 82, 88, 90, 98, 107, 119
China, 88, 89, 95
chitin, 6, 10, 14, 17, 19
chloride, 5
chloroform, 162, 165, 166
cholinergic, 38
chromatography, 30
chromosome, xi, 31, 161, 166, 167
chromosomes, xi, 161, 162, 163, 167
chronic, 106, 107, 110
classes, 30
classical, 25
classification, viii, 77, 78, 116, 152, 159
cleaning, 42, 57
cleavage, 4, 9, 19, 24
climate change, 79, 100
clinical, x, 103, 106, 107, 108, 109, 110, 111, 112, 113, 114, 115, 116, 117, 118, 120
clinical symptoms, viii, 59, 111, 113
clinical syndrome, 109
clinical trial, 116
clinical trials, 116
clinically significant, 107
clone, 50
cloning, 28, 50, 119, 166, 168, 169
closure, 38
cluster of differentiation (CD), 99
clusters, 31
CMV, 111, 173
Co, 72, 74, 124, 129, 174

coagulation, 27, 36
cockroach, 36
coding, 4, 9, 56
codon, 113
co-existence, 48
cohort, 124, 125, 126, 137
Coleoptera, 2, 17
collagen, 96
Colony Collapse Disorder, 49
Colorado, 78, 81, 98, 101, 147
Colorado tick fever, 78, 81, 98
Columbia, 81, 151
coma, 116
common signs, 90
communication, 71
communities, 122
community, 95, 122, 141, 143
competence, 96, 136, 140, 143
complement, 30, 91
complex interactions, 30
complications, 80, 89, 119
components, 19, 23, 26, 35, 71
composition, 9, 32, 113
compounds, 11
comprehension, 112
computed tomography (CT), 91, 106
concentration, x, 106, 108, 109, 110, 111, 112, 128, 152, 153, 155, 157, 162, 165
conduction, 109, 113
confusion, 112
Congress, iv, 142
conjunctivitis, 105
Connecticut, 99, 122, 143, 144
consciousness, 84, 106
conservation, 60, 71
contamination, 50, 162
control, vii, xi, 1, 2, 10, 13, 20, 23, 26, 28, 57, 60, 107, 116, 129, 130, 137, 152, 153, 157, 159, 166, 171
control group, 130, 137
controlled, 115, 120
controlled studies, 115, 138
convergence, 145
convulsion, 90
convulsions, 84, 90
corn, 19
correlation, 47, 107, 113, 114
cortex, 91
cortical, 106
costs, 2, 10, 12

cranial nerve, 111, 112
crops, 10, 12, 43, 74, 152
cross-sectional, 137
crustaceans, 23
crystalline, 152
C-terminal, 19
C-terminus, 9
culex, ix, 103, 104
cultivation, 79
culture, 4, 12, 31, 56, 108, 109, 111, 131, 144, 157, 173
culture media, 173
cuticle, 24
cycles, x, 62, 104, 121, 143, 144
Cyprus, 88
cysteine, 10
cysteine residues, 10
cytokine, 26, 38
cytoplasm, 3, 167
cytoskeleton, 30
Czech Republic, 87

D

D. melanogaster, 31, 168
Dallas, 36, 82
danger, ix, 60, 78, 79
data analysis, 173
database, 7
de novo, 167
death, 10, 48, 90, 91, 115
deaths, 82, 84
defects, 35, 111, 136
defense, viii, 19, 27, 33, 35, 36, 37, 41, 43, 49, 53, 54, 114, 115, 136, 172
defense mechanisms, 136
defenses, 57
deficiency, 26, 107, 114, 115
deficit, 47
deficits, 47, 55, 111
definition, 106
deforestation, ix, 77, 79, 82
degradation, 9, 19
degrading, vii, 21
degree, 108, 110, 114, 115, 116
delivery, 152
demographic change, ix, 77
demyelination, 109, 111
denervation, 110
dengue, 80, 84, 92, 99, 100, 101, 109, 119

dengue fever, 84, 100, 109
density, 24, 26, 31, 127, 143, 164, 165, 167, 172
Department of Agriculture, 151
depressed, 112
derivatives, 11
detection, viii, xi, 54, 55, 59, 62, 63, 65, 68, 71, 72, 73, 74, 75, 90, 91, 95, 98, 106, 118, 131, 133, 144, 146, 157, 162, 166
developed countries, 109
diabetes, 107, 114
diabetes mellitus, 107, 114
diagnosis, ix, 78, 80, 91
diagnostic, 106, 107, 116, 119
diagnostic criteria, 119
diamond, 6, 14
diarrhea, 105
diet, x, 5, 60, 151, 153, 155, 157
differential diagnosis, 90
differentiation, 94, 101
diffusion, ix, 77, 79, 80, 83, 84, 85, 87, 88, 89, 91, 92
disability, 82, 88, 90, 115
diseases, 57, 72, 79, 80, 81, 96, 97, 98, 113, 114, 139, 141, 142
Disease-specific, 143
disorder, 53, 73, 111
dissociation, 111
distal, 110
distress, 109, 110
distribution, ix, 51, 53, 69, 70, 72, 77, 79, 100, 104, 107, 125, 126
diversity, 38, 56, 58, 60, 74, 82, 95
division, 42, 58, 171
division of labor, 42, 58
DNA, xi, 2, 6, 7, 32, 54, 101, 115, 122, 124, 131, 133, 144, 146, 152, 159, 161, 162, 163, 164, 165, 171, 174
DNA sequencing, 124
dogs, 122
dominance, 54
dosage, 155
draft, 51
Drosophila, vi, xi, 22, 23, 24, 26, 27, 29, 30, 31, 32, 34, 35, 36, 37, 38, 46, 50, 54, 55, 56, 161, 162, 165, 166, 167, 168, 169
drugs, 23
dry ice, 71
duration, 88, 113, 114, 130
dysphagia, 109

E

E. coli, 13, 26, 28, 29, 30, 31, 32, 50
earth, vii
East Asia, ix, 47, 78, 79
Eastern Europe, 79, 83, 87
E-coli, 37
ecological, viii, 77, 78, 86, 100
ecology, ix, 50, 60, 77, 79, 97
ecosystems, 60
edema, 115
EEG, 91, 106
efficacy, 113, 115, 116
egg, 37
Egypt, 84, 85, 87, 96, 101, 120
ehrlichiosis, 136, 137, 139, 140, 141, 142, 143, 144, 145, 146, 147
elderly, 107, 110
electron, 8, 17
electron microscopy, 8
electrophoresis, xi, 62, 161, 173
electrophysiologic, 109, 110
ELISA, 91, 92, 93, 136, 137
embryo, 93, 113
embryos, 33
EMG, 109, 110
emigration, 136
employment, 23, 153
encapsulated, 24
encapsulation, viii, 21, 24, 28, 29, 30, 36, 38
encephalitis, viii, ix, x, 77, 78, 79, 80, 81, 82, 83, 84, 85, 86, 87, 88, 89, 90, 92, 94, 95, 96, 97, 98, 99, 100, 101, 102, 103, 104, 106, 107, 108, 109, 110, 112, 114, 115, 116, 117, 118, 119, 120, 122, 123, 128, 142
encephalomyelitis, 78, 90, 97, 118
encoding, 9, 14, 15, 28
endemic, 80, 82, 86, 88, 89, 92
endocytosis, 24
endoplasmic reticulum, 3, 4, 16
endothelial cell, 91, 113
endothelial cells, 91, 113
endothelium, 114
England, 86, 87, 89
enteric, 89
enterovirus, 112, 119
enteroviruses, 106, 109, 112
entrapment, 27, 36
envelope, 104

environment, vii, ix, xi, 1, 2, 10, 11, 57, 78, 79, 152, 171, 172
environmental change, 172
enzymatic, 8
enzyme, 106, 120
enzyme immunoassay (EIA), 106, 109, 120
enzyme-linked immunosorbent assay (ELISA), 91, 92, 93, 137
epidemic, ix, 78, 79, 80, 81, 82, 86, 88, 90, 96, 98, 101, 106, 107, 108, 109, 110, 112, 117, 118
epidemics, ix, 81, 82, 84, 103, 105, 106, 108, 110, 115
epidemiology, viii, ix, 41, 77, 78, 79, 85, 92, 94, 102
epithelia, 9, 15
epithelial cell, 25, 152
epithelial cells, 25, 152
epithelium, vii, 1, 2, 20
epitopes, 111, 114
Epstein-Barr virus, 111, 119
equilibrium, 164, 165, 167
erosion, 60
Escherichia coli, 12, 25, 30, 34
esterase, 19
ethanol, 124, 165
ethyl alcohol, 5
etiologic agent, 122, 139
etiology, 91, 110, 138
etiopathogenesis, 90
eukaryotic cell, 172
Euro, 81, 82, 85, 86, 87
Europe, viii, ix, 46, 59, 69, 74, 79, 83, 84, 85, 86, 87, 92, 98, 103, 117, 122, 139, 143
European Union, 60
euthanasia, 131
Everglades, 82
evidence, x, 107, 114, 115, 121, 123, 128, 135, 138, 139, 143, 144, 145, 146, 147
evolution, ix, x, 24, 36, 48, 49, 55, 78, 79, 90, 103
excision, 95
exotic, 118
experimental condition, 162
exposure, x, 61, 104, 107, 120, 121, 137, 143, 152, 153, 154, 155
expressed sequence tag (EST), 21, 50
external environment, ix, 78, 79
extinction, ix, 78, 79
extraction, xi, 62, 124, 159, 162, 165
eye, 111

F

facial nerve, 110, 112
failure, 27, 110, 112, 138
false negative, 71
family, ix, xi, 3, 7, 25, 30, 31, 34, 37, 54, 55, 57, 79, 81, 88, 99, 103, 104, 115, 122, 152, 166, 168, 171
FAO, 60
Far East, 88, 89, 98
farming, 79, 99
farms, 48
fat, 3, 25, 26, 174
fatalities, 107
fauna, 95
fear, 82
febrile, 84
feeding, x, 8, 12, 15, 16, 38, 42, 112, 115, 123, 124, 125, 126, 127, 129, 131, 132, 133, 134, 136, 144, 151, 153, 155, 157
feet, 110
females, xi, 161, 163, 166
fetal, 173
fever, ix, 78, 79, 80, 81, 82, 84, 88, 89, 90, 92, 95, 96, 98, 99, 100, 101, 102, 103, 105, 106, 108, 109, 112, 117, 118, 119, 120, 122, 139, 141, 143, 145, 146, 147
fibers, 109
fibrillation, 109
fibroblasts, 94, 113
fire, 50, 57
fitness, 42, 122
fixation, 91
flaviviruses, 82, 96, 97, 101, 109, 112, 113, 115, 116
fluid, 60, 106
fluorescence (FL), x, 151, 153, 154, 155, 156, 157
focusing, 32, 47
food, viii, 41, 42
forceps, 62
forestry, vii, xi, 1, 2, 10, 171
forests, 87, 142
formaldehyde, 5
fractionation, 164, 165
France, 57, 75, 86
functional analysis, 25
fungal, 27, 33, 136
fungi, 23, 24, 25, 26, 28, 29, 43
fungus, 27
fusion, 6, 8, 9, 16, 20

G

ganglia, 90
gel, xi, 161, 162, 163, 164, 165, 166, 173
gels, 62, 124, 173
gene, vii, x, xi, 2, 4, 8, 9, 10, 13, 14, 15, 16, 17, 18, 19, 21, 23, 24, 26, 27, 29, 30, 31, 33, 35, 36, 37, 38, 43, 62, 63, 101, 113, 119, 124, 151, 152, 153, 157, 159, 160, 161, 166, 167, 168, 171
gene expression, 26, 27, 43, 159, 171
gene silencing, vii, 21, 23, 33
gene transfer, 159
generalization, xi, 171
generalized seizures, 112
generation, 33, 167
genes, xi, 4, 8, 9, 10, 12, 15, 18, 23, 24, 26, 27, 28, 30, 31, 32, 33, 34, 36, 38, 43, 49, 55, 57, 152, 161, 166, 167, 172
genetic information, 55
genetically modified organisms, 10
genetics, 22, 24, 32, 36, 37, 42, 55, 57
genome, xi, 8, 10, 14, 17, 21, 26, 30, 31, 32, 35, 38, 43, 44, 46, 48, 49, 50, 52, 53, 54, 55, 56, 57, 58, 73, 74, 113, 117, 152, 161, 162, 165, 166, 167, 168, 171
genome sequences, 43, 117, 166
genome sequencing, 33, 43, 49, 54
genomes, 46, 48, 50
genomic, xi, 14, 32, 37, 43, 50, 57, 58, 60, 61, 95, 161, 162, 163, 166
genomics, 36
genotype, 98, 162, 167
genotypes, 53, 163
genus, viii, 77, 78, 79, 81, 88
Germany, 62, 86, 87, 92, 98, 100, 139, 141, 142, 143
Gibbs, 49, 53, 72
gland, 33, 35, 55
glial, 91, 114
glial cells, 91, 114
glucose, 106, 108, 109, 111
glycoprotein, 4, 15
glycoproteins, 4, 6
glycosylated, 4
glycosylation, 4
goat milk, 87
government, iv
Gram-negative, 26, 29, 30
gram-negative bacteria, 26
Gram-positive, 29
gram-positive bacteria, 26
granules, 4, 5, 6, 8, 9, 11, 12
Greece, 92
green fluorescent protein, 171
groups, viii, 2, 41, 43, 78, 93, 94, 105, 107, 123, 127, 129, 141
growth, 26, 91, 172, 174
growth factor, 26
guidelines, 124
Guillain-Barre syndrome, x, 103, 109, 110
Guinea, 88, 89, 101
Gulf of Mexico, 83
gut, 25, 27, 104

H

habitat, 79, 86
haemostasis, 27
Haifa, 103
hands, 110, 111
Harvard, 58
head, 112
head injury, 113
headache, ix, 88, 89, 90, 92, 103, 105, 108, 112
healing, 27, 38
health, xi, 60, 61, 73, 95, 97, 100, 117, 146, 171
health problems, 97, 146
health status, 73
heat, x, 146, 151, 153
heating, 5
hematologic, 100
Hemiptera, 33, 55
hemisphere, 112, 113
hemolymph, 25, 26, 37
hemorrhages, 106
hepatitis, 84, 85, 105, 115, 116
hepatitis a, 85
hepatitis C, 115, 116
hepatosplenomegaly, 105
herpes, 90, 111, 112
herpes simplex, 90, 112
herpes virus, 111
Higgs, 101
high risk, 93, 94, 116, 137
high temperature, 3, 5, 10
high-risk, 141, 142
histamine, 33
HIV, 107, 114, 115
hives, 57, 69
homolog, 27
homologous genes, 36

homology, xi, 46, 171
Honda, 89, 97
honey, 24, 27, 33, 42, 43, 52, 53, 54, 55, 56, 57, 58, 60, 61, 63, 65, 69, 72, 73, 74, 75
honey bees, 53, 55, 57, 58, 63, 65, 72, 73, 74, 75
horses, ix, 81, 86, 88, 101, 102, 103, 108, 116, 118, 120, 122
hospital, 104, 106, 107, 110, 118
hospitalization, 111, 116
hospitalized, 97, 105, 106, 112, 117
host, v, vii, viii, ix, x, xi, 1, 3, 4, 5, 6, 7, 10, 11, 12, 13, 14, 16, 17, 20, 21, 23, 28, 31, 33, 35, 36, 41, 46, 48, 49, 51, 52, 56, 57, 58, 78, 79, 94, 97, 99, 101, 104, 111, 113, 114, 115, 116, 121, 122, 123, 125, 136, 137, 138, 140, 144, 152, 162, 166, 167, 171, 172, 174
household, 106, 117
human, viii, x, 26, 31, 60, 77, 78, 81, 86, 91, 92, 93, 94, 95, 96, 97, 104, 106, 113, 115, 117, 118, 120, 121, 122, 128, 131, 137, 139, 140, 142, 143, 144, 145, 146, 147
human genome, 31, 113
humane, 124
humans, vii, x, 23, 30, 42, 60, 84, 86, 97, 98, 104, 105, 108, 114, 115, 120, 121, 122, 123, 128, 136, 138, 162
humidity, 124, 130
humoral immunity, 115
Hungarian, 72
hybridization, xi, 101, 162
hybrids, xi, 101, 161, 163, 166
hydrolysis, 25
hydrophobic, 10
hygienic, 49, 57
hypertension, 90, 107, 114
hypokalemia, 105
hyporeflexia, 108
hypothesis, 104, 115, 172

I

ice, 71, 129, 131, 133
id, 10, 138
identification, xi, 25, 29, 37, 43, 50, 55, 72, 73, 99, 101, 146, 162, 166, 167, 168
identity, 26
IgG, 114, 120, 131, 132, 133
image analysis, 157, 159
imaging, 90, 91, 106, 113
immune cells, 32
immune function, 28
immune response, viii, 21, 23, 26, 28, 29, 30, 31, 32, 34, 35, 37, 38, 47, 49, 104, 111, 114, 135, 136, 138, 139, 146, 162, 168
immune system, 23, 24, 28, 29, 37, 38, 49, 58, 90, 100, 113, 136
immunity, vii, viii, ix, 21, 23, 24, 26, 27, 28, 30, 32, 34, 35, 36, 37, 38, 48, 49, 51, 78, 79, 93, 114, 115, 116, 168, 169
immunization, 116, 137
immunoassays, 120
immunocompromised, 107, 114
immunofluorescence, 91, 124, 131, 144
immunoglobulin, 34, 106, 116, 120
immunoglobulins, 115
immunohistochemical, 107, 113, 118
immunohistochemistry, 97
immunological, 97, 138
immunology, 49, 117
immunosuppression, 136
in situ, xi, 162
in situ hybridization, xi, 162
in utero, 104
in vitro, 7, 12, 28, 34, 94, 120, 145
in vivo, 6, 15, 26, 28, 31, 34, 35, 115, 145
inactivated polio vaccine (IPV), 109
inactivation, 24, 26, 82, 95, 152
incidence, 61, 65, 72, 82, 88
inclusion, 2, 3, 5, 11, 14, 15, 16, 116
inclusion bodies, 3, 5, 11, 14, 15, 16
incubation, 89, 93, 94, 105
incubation period, 89, 105
India, 88, 89, 96, 100
Indian, 14, 84
indicators, 100
indices, 106, 108
Indonesia, 89, 119
induction, 26, 27, 47
industrial, 60
inert, 32
infections, vii, viii, ix, x, 1, 6, 16, 21, 26, 27, 32, 43, 47, 59, 75, 77, 78, 79, 80, 81, 89, 90, 91, 92, 95, 97, 98, 99, 100, 101, 102, 103, 104, 106, 107, 111, 113, 115, 116, 117, 118, 119, 121, 122, 128, 129, 132, 134, 135, 136, 137, 138, 140, 141, 142, 145
infectious, viii, 5, 18, 41, 48, 50, 53, 55, 56, 100, 111, 112, 117, 120, 125, 131, 135, 137, 141
infectious disease, viii, 41, 100, 120, 141
infectious diseases, 100, 120, 141

infestations, 128, 131, 132, 133
inflammation, 105, 107, 111, 112
inflammatory, 105, 107, 110, 111, 114, 115
influence, 122
influenza, 89, 112, 119
ingestion, 6, 33, 80
inhalation, 84
inhibition, 91, 102, 113, 147, 166
inhibitor, 26
initiation, 26, 115, 116
injection, 24, 29, 33, 37, 42
injections, 27
injury, iv, 24, 27, 113
innate immunity, 23, 24, 26, 27, 30, 32, 34, 35, 37, 38, 169
inoculation, 5, 12, 135, 136
inoculum, 107
insecticide, 10
insecticides, vii, 1, 2, 12, 13, 61, 153
insects, vii, 1, 2, 6, 7, 9, 10, 12, 15, 18, 19, 20, 22, 23, 24, 25, 26, 27, 28, 29, 30, 31, 32, 35, 37, 38, 42, 43, 48, 49, 50, 54, 55, 56, 57, 74, 153, 162, 166
insertion, 112
insults, 112
integration, 8
integrin, 29, 36
integrity, 15, 71, 73, 141
integument, 33
interaction, viii, x, 23, 41, 50, 121, 122, 138
interactions, xi, 30, 34, 49, 51, 54, 57, 58, 113, 122, 136, 144, 162, 167
interface, 144
interference, vii, x, 21, 29, 31, 33, 34, 35, 36, 37, 38, 121, 122, 166, 169
interferon (IFN), 93, 94, 116, 120, 172
internal ribosome entry site, 58
interrelationships, 122
interval, 125, 126, 127, 135
intestine, 46, 57
intravenous (IV), 89, 111, 120
intrinsic, 28, 51
invading organisms, 24
invertebrate, 80
invertebrates, 26, 34
Ireland, 86, 87, 89
irritation, 108
island, 120
isoforms, 29, 34
isolation, 91, 95, 97, 106, 119, 145, 146

Israel, ix, 56, 71, 74, 84, 86, 87, 103, 104, 108, 116, 117
Italy, 77, 86, 87, 88, 92, 94, 97, 100, 101, 102, 118, 140

J

JAMA, 100, 145
Jamestown, 79, 81, 83, 99
Japan, 1, 17, 41, 46, 47, 48, 51, 55, 88, 89, 98, 99
Japanese, ix, 43, 48, 78, 79, 80, 88, 89, 90, 95, 96, 98, 100, 101, 104, 106, 107, 109, 112, 115, 116, 117, 119, 120
Japanese encephalitis, ix, 78, 79, 80, 88, 89, 90, 95, 96, 98, 100, 101, 104, 106, 107, 109, 112, 115, 116, 117, 119, 120
Java, 62, 102
JNK, 26
Jordan, 120
Jun, 26

K

Kashmir, 45, 60, 73, 75
Kenya, 85
killing, xi, 10, 12, 25, 27, 37, 171
kinase, 26
kinetics, 29
King, 3, 15, 16

L

L2, 165
labor, 10, 12, 42, 53, 58
labor-intensive, 10, 12
language, 112
large-scale, 5
larva, 5, 153, 154, 155
larvae, x, 3, 5, 6, 8, 9, 11, 12, 13, 14, 15, 16, 17, 18, 19, 20, 27, 28, 30, 31, 42, 48, 49, 60, 124, 127, 128, 129, 130, 131, 132, 133, 134, 139, 151, 152, 153, 154, 155, 156, 157, 159
larval, 19, 31, 35, 123, 124, 126, 127, 129, 134, 135, 143, 153
latex, 30, 36
Latin America, 81, 82, 83, 84
learning, 47, 55, 56
lecithin, 11, 19
left hemisphere, 112, 113

Lepidoptera, 2, 3, 7, 14, 15, 16, 17, 18, 19, 20, 35, 174
lepidopterans, 23
lesions, 90, 99, 106, 112, 114, 144
leucocyte, 147
leukocyte, 114
leukocytosis, 105
leukopenia, 105
life cycle, 60
life forms, vii
life style, 60
lifespan, 49
life-threatening, 115
ligand, 30
lignin, 12
limb weakness, 110
linear, 2, 153, 165
linear regression, 153
lipid, 34
lipid metabolism, 34
lipopolysaccharide, 29, 36, 38
lipoprotein, 31
liquid nitrogen, 71, 124
liquor, 91
Listeria monocytogenes, 112, 119, 135
literature, 105, 119
liver, 105
livestock, 10, 99, 136
localization, ix, 78, 79, 167
location, 105, 113
locus, 31, 56, 167, 171
London, 16, 56, 62, 116
long period, 48
longitudinal study, 96
long-term, 111, 114
losses, 61
Louisiana, 82, 97, 109, 119
low risk, 92, 93
luciferase, 157
lumbar, 108, 110
lumbar puncture, 108
lumen, 6, 152
lung, 107
lung disease, 107
lymph, ix, 103, 105, 113
lymph node, 113
lymphadenopathy, ix, 103, 105
lymphocyte, 109, 136, 147
lymphocytes, 114
lymphoma, 107, 112, 118
lysis, 162
lysozyme, 25, 29

M

machinery, 32
macrophage, 30
macrophages, 113
magnetic, iv, 90, 91, 107
magnetic resonance imaging (MRI), 90, 91, 107, 109, 112, 113
maintenance, viii, 41, 80, 123, 124, 135
malaise, 105
malaria, 36
males, xi, 161, 163, 166
malignant, 114
mammalian cell, 159, 172
mammalian cells, 159, 172
mammals, 23, 42, 80, 81, 84, 86, 98, 120, 139, 143
manipulation, 53
mannitol, 115
manufacturer, 62
Maori, 45, 49, 56, 71, 74
marital status, 151
Mars, 98
masking, 138
mass spectrometry, 173
matrix, 3, 8, 15, 18, 19, 33, 57, 124, 152
matrix protein, 19
Mauritania, 85
mean, 88, 89, 90, 93
mechanical, 109, 110, 115
mechanical ventilation, 109, 110, 115
media, 172, 173
mediation, 25
medication, 113
medications, 113, 115
Mediterranean, 84, 85, 86, 87, 88
medulla, 107, 112, 114
medulla oblongata, 112
melanin, 24, 28, 39
memory, 28, 47
memory deficits, 47
men, ix, 77, 79, 80, 85, 87
meninges, 90, 105
meningitis, x, 80, 81, 82, 88, 90, 100, 103, 105, 108, 109, 110, 118, 139
meningoencephalitis, x, 88, 90, 103, 105, 106, 108, 110, 111, 112, 118, 119, 120
mental status change, 108, 110

mental status changes, 110
messenger RNA, 24
metabolism, 34, 55, 172
metalloproteinase, 172
metazoan, 24
metazoans, 30
Mexico, 81, 83
Miami, 171
mice, x, 113, 121, 122, 123, 124, 125, 126, 127, 128, 129, 130, 131, 132, 133, 134, 135, 136, 138, 144
microaggregates, 28
microbes, 49, 53, 73, 96
microbial, 23, 26, 28, 35, 43, 49, 116
microglia, 91
microglial, 107, 114
microorganisms, 25, 137
microscope, x, 151, 153, 155
microscopy, 8, 16, 142
midbrain, 112
Middle East, 86, 87, 88, 117
migration, 102, 147
migratory birds, 104
milk, 80, 87
mimicry, 111
Minnesota, 144
MIP, 28
Mississippi, 109, 119
Missouri, 151
mites, 46, 47, 58, 72, 74
mitochondrial, 165
mitochondrial DNA, 165
mitogenesis, 136
mobility, 137, 139
modality, 83, 84
model system, 30
modeling, 56
models, 99
modulation, viii, 21
molecular biology, 14, 22, 50, 152
molecular mass, 25, 31
molecular mimicry, 111
molecular structure, 12, 26
molecular weight, xi, 161, 162, 163, 165
molecules, 25, 27
molting, 36, 130, 137, 157
monkeys, 115, 120
Monoclonal antibodies, 29
monocytes, 113
monsoon, 88
Moon, 45, 56

morbidity, ix, 103, 105, 107, 112
morphogenesis, 38
morphological, xi, 161, 166, 168
morphology, 166
mortality, viii, ix, x, 12, 59, 60, 61, 63, 64, 65, 66, 67, 68, 69, 70, 71, 72, 81, 82, 88, 90, 103, 105, 106, 107, 112, 152, 153, 155, 157
mortality rate, 12, 81, 82, 88, 90, 105, 106, 107
Moscow, 141, 143, 161
mosquitoes, ix, 31, 32, 77, 79, 81, 82, 83, 84, 85, 86, 88, 97, 98, 99, 103, 104, 107, 116, 118, 120
mothers, 104
moths, 31
mouse, 100, 113, 123, 124, 127, 129, 130, 133, 135
mouth, 56
mRNA, vii, 21, 24, 27, 28
mucin, 9, 19, 27, 35
multiple sclerosis, 111
multiplicity, 173
muscle, 108, 109, 110, 111
muscle strength, 108, 110
muscle weakness, 108, 110, 111
muscles, 110, 111
mutant, 6, 10, 12, 37
mutants, 24, 56
mutation, 31
mutations, 46
myalgia, ix, 103, 105
mycobacterial infection, 30, 37
Mycobacterium, 30
myocarditis, 105

N

nanometers, 165
Nash, 108, 117
nasogastric tube, 112
national, 116
National Academy of Sciences, 19, 56, 57
National Institutes of Health, 124
national origin, 151
native population, 102
natural, viii, x, xi, 4, 8, 11, 24, 26, 35, 41, 42, 43, 48, 60, 61, 74, 98, 100, 108, 117, 121, 122, 123, 128, 135, 136, 137, 138, 143, 167, 168, 171
natural enemies, 42
natural environment, 60, 61
natural selection, 24
nausea, 105
NCS, 90

neck, 108
necrosis, 113
needle aspiration, 145
negative regulatory, 39
nematodes, 23, 24, 35, 38
Nepal, 46
nephritis, 85
nerve, 106, 109, 110, 111, 112, 114
nerve conduction velocity (NCV), 109, 110
nerves, 112
nervous system, ix, 86, 96, 97, 99, 100, 103, 104, 105, 111, 113, 118
neural tissue, 113, 114
neuritis, x, 103, 111, 112, 119
neuroimaging, 110
neurologic, 81, 82, 84, 88, 89, 90, 92, 99, 100
neurologic symptom, 105
neurological disability, 90
neurological disease, 81, 106
neurological disorder, 96
neuronal cells, 90, 91, 114
neuronal degeneration, 107
neuronal plasticity, 88
neurons, 38, 97, 113
neuropathological, 118
neuropathology, 119
neurophysiology, 42, 55
neuroscience, 49
neutralization, 91
neutropenia, 136
neutrophil, 136, 141
New Jersey, 106, 117, 146
New South Wales, 96
New World, 95
New York, iii, iv, ix, 15, 16, 56, 83, 103, 104, 106, 107, 108, 109, 110, 112, 115, 117, 118, 120, 137, 140, 145, 159, 174
next generation, 33
Nicotiana tabacum, 13, 16
Nigeria, 85
Nile, v, ix, 78, 79, 81, 83, 84, 85, 86, 95, 96, 97, 99, 102, 103, 104, 105, 117, 118, 119, 120
nitrogen, 71, 124
nodes, 113
nodules, 25, 28, 107, 114
normal, 48, 50, 106, 108, 109, 110, 111
North Africa, 84
North America, viii, 59, 69, 81, 82, 84, 86, 117, 122
North Carolina, 82, 101
Northeast, 82

Norway, 139
N-terminal, 9, 10, 19, 26
nuchal rigidity, 88, 110
nuclear, 13, 14, 15, 16, 19, 20, 34, 159, 160, 174
nuclei, 91, 112, 152
nucleic acid, xi, 62, 106, 113, 161, 162, 163, 164, 165, 167
nucleocapsids, 6, 8
nucleotide sequence, 15, 17, 48, 53, 54, 56, 73, 124, 146
nucleotides, 104
nucleus, 7
nurse, 42, 44
nymphs, x, 121, 123, 124, 125, 126, 127, 128, 129, 130, 131, 132, 133, 134, 135, 139, 142

O

OB, x, 152, 153, 154, 155, 156, 157, 158
observations, 17, 101, 117, 119, 136
occlusion, x, 4, 5, 6, 7, 13, 15, 16, 18, 19, 48, 151, 152, 154, 156, 159
occupational, 107, 115
occupational therapy, 107, 115
Oceania, 69
ocular, 101
Ohio, 171
oils, 12
old age, 107
olfactory, 91
online, 62
operon, 146
optic disc, 111
optic nerve, 112
optic neuritis, x, 103, 111, 112, 119
optical, 11, 12, 14, 18
optimization, 71, 142
oral, 8, 10, 13, 16, 109, 112
oral polio vaccine, 109
organ, 104, 113, 122
organism, viii, 23, 41, 43, 122
outpatient, 107
oxygen, 25

P

pain, 105, 111
pancreatitis, 105, 117
Papua New Guinea, 88, 101

paralysis, viii, x, 43, 45, 49, 53, 54, 56, 57, 58, 59, 60, 71, 72, 73, 74, 90, 103, 108, 109, 110, 118, 119, 160
parasite, 25, 26, 32, 35, 45, 49, 74, 122, 144
parasites, 14, 24, 25, 26, 30, 37, 52, 56, 57, 74, 122, 137
parenchyma, 105, 114, 115
parents, xi, 161, 166
paresis, 108
paresthesias, 109
particles, xi, 14, 113, 152, 157, 162, 166
passive, 91
pathogen, 85
pathogenesis, 14, 15, 20, 32, 58, 104, 110, 118
pathogenic, viii, 28, 34, 43, 45, 48, 51, 52, 58, 59, 84, 122, 136, 137
pathogenic agents, 122
pathogens, viii, x, 24, 28, 29, 30, 35, 41, 43, 49, 56, 59, 60, 61, 69, 111, 112, 117, 121, 122, 123, 124, 125, 127, 128, 129, 130, 132, 133, 135, 136, 137, 138, 140, 143, 144, 146
pathology, viii, 52, 77, 78, 118
pathophysiology, 117
pathways, 23, 26, 29, 30, 32, 34, 36, 55
patients, 81, 82, 88, 90, 97, 105, 106, 107, 108, 109, 110, 111, 112, 113, 114, 115, 116, 117, 128, 137, 146, 147
pattern recognition, viii, 21, 27, 28
pediatric, 107
penalties, 57
peptide, 4, 26, 37, 173
peptides, viii, 21, 24, 25
peripheral blood, 147
peripheral nervous system, 111
perivascular cuffing, 107, 114
periventricular, 107
permeability, 9, 18
permeabilization, 25
permit, 91
personal communication, 71
personality, 88, 90
Peru, 81
pest control, vii, 1, 2, 10, 159
pesticides, vii, 1, 2, 5, 10, 12, 17, 61
pests, xi, 10, 12, 152, 171
Petri dish, 153
pH, 16, 131
phagocytic, 31, 141
phagocytosis, viii, 21, 24, 25, 28, 30, 31, 32, 34, 36, 37, 136

pharyngitis, 89
phenol, 33, 162, 165, 166
phenotype, 24, 31, 37, 47
phenotypes, 5
phenotypic, 24
Philadelphia, 118, 144
phosphate, 62, 131
phospholipids, 8
photophobia, 90
phylogenetic, 46, 73, 117
phylogeny, 54
phylum, 24
physical therapy, 107, 115
Physicians, 144
physiological, 157
physiology, 18, 22, 137
pigs, 88, 96, 135
placebo, 115, 120
plants, vii, 1, 2, 10, 12, 13, 16, 21, 23, 34, 54, 60, 152
plaque, 106, 172
plaques, 172
plasma, 24
plasmapheresis, 111
plasmids, 50
plastic, 62, 124, 153
plasticity, 88
play, 3, 6, 12, 60, 61, 106, 113, 136
pneumonia, 111, 112
Poland, 87, 101, 146
polio, 108, 109, 110
poliovirus, 43, 46, 55, 56, 118
pollen, 42, 60
pollination, 60
pollinators, 74
polymerase, 62, 63, 146
polymerase chain reaction (PCR), 44, 53, 54, 55, 58, 61, 62, 71, 72, 73, 74, 75, 91, 106, 107, 111, 113, 115, 124, 125, 126, 127, 130, 131, 132, 133, 140, 144, 146
polymerization, 37
polymorphism, 167, 168
polymorphonuclear, 108, 147
polymorphonuclear cells, 147
polypeptides, 56, 168
polystyrene, 32
pools, 124, 127
population, viii, ix, xi, 41, 58, 61, 78, 79, 80, 84, 89, 92, 115, 123, 140, 142, 153, 161, 162, 163, 167
port of entry, 113

Portugal, 86, 87, 88
power, 29
Poxviridae, vii, 1, 2
PPO, 24, 27, 29
predators, viii, 41, 42
preference, 97
preparation, 116
prevention, 116
probability, 125, 126, 137, 142
procedures, 109
production, xi, 2, 5, 10, 12, 13, 17, 18, 25, 60, 61, 91, 93, 94, 114, 136, 152, 161, 166, 167
production costs, 2, 10, 12
prognosis, 105, 107, 110, 112
prognostic, 100
prognostic factors, 110
program, 45, 62
progressive, 111
proliferating fraction (PF), 99
proliferation, 91, 122, 136
promote, 114
promoter, x, 26, 151, 157, 171
property, iv, 19
prophylaxis, ix, 78
proteases, 35
protection, viii, 3, 41, 57, 160
protective role, 116
protein, x, 2, 3, 4, 5, 7, 8, 9, 10, 12, 13, 14, 15, 16, 17, 18, 19, 20, 25, 26, 27, 28, 29, 30, 31, 33, 34, 35, 36, 37, 38, 39, 42, 101, 104, 106, 108, 109, 110, 111, 112, 113, 114, 151, 152, 153, 157, 159, 168, 171
proteins, v, vii, viii, 1, 2, 6, 8, 9, 10, 12, 13, 14, 16, 19, 21, 25, 26, 28, 29, 30, 31, 33, 34, 35, 55, 57, 104, 136, 142, 152, 172, 173
proteoglycans, 6
protocol, 165
protocols, 165
protozoan parasites, 26
proximal, 114
public, 116, 117
public education, 116
public health, xi, 95, 97, 117, 171
pupae, 27, 33, 42, 60
pupal, 31, 36
purification, xi, 12, 28, 50, 162, 165, 167

Q

Quantitative trait loci, 55

quinones, 28

R

race, 23, 151
radiation, 3, 5, 10
radiological, 98
rain, 52, 57, 107
rainfall, 82, 84, 88, 98
range, vii, xi, 1, 2, 10, 48, 56, 65, 104, 105, 106, 109, 115, 152, 171, 172, 174
RAPD, 111
RAS, 167
rash, ix, 103, 105
rat, 143
rats, 135
reactive oxygen, 25
reactive oxygen species, 25
reactivity, 136, 141, 144
reality, 35, 85, 90
recall, 80, 84
receptors, 26, 31, 109
reciprocal cross, xi, 161, 163, 166
recognition, viii, 21, 23, 25, 26, 27, 28, 29, 30, 34, 38
recombination, 56, 74, 82, 145, 172
recovery, 108, 110, 112, 114, 115
reduction, 106
reflection, 116, 156, 157
reflexes, 108, 109, 111
refractoriness, 32
regional, 113
regression, 153, 158
regression equation, 153
regular, 165
regulation, 26, 27, 28, 30
regulators, 34
relationship, 43, 46, 48, 49, 50, 51, 56, 61, 72
relationships, 14, 55, 95, 122, 137, 140
religion, 151
remission, 89
replication, vii, ix, xi, 1, 2, 4, 6, 8, 14, 17, 31, 90, 93, 94, 97, 103, 104, 113, 114, 115, 120, 157, 161, 166, 167, 171, 172, 173
representative samples, 124
reproduction, viii, 41, 42
reptiles, 80
research, 106, 116
researchers, 114

reservoir, ix, 78, 79, 84, 88, 94, 99, 100, 103, 122, 123, 137, 138, 143
reservoirs, 79, 80, 81, 83, 85, 86, 87, 88, 89, 93, 136, 144
residues, 4, 10
resistance, 20, 24, 26, 28, 57, 58, 99, 113, 119, 166, 168
resolution, 108, 115, 165
respiratory, 86, 109, 110, 111
respiratory failure, 110
response, 93
retention, 109
reticulum, 3, 4, 16
retroviruses, 167
reverse transcriptase, 118, 168
Reynolds, 32, 34, 37
ribosome, 58
rice, 13, 16, 79, 143
Rift Valley fever, 80, 84, 95, 98, 99, 101
rigidity, 88, 90, 110
RISC, 24
risk, ix, 78, 79, 80, 84, 89, 91, 92, 93, 94, 107, 112, 114, 115, 116, 134, 137, 141, 142, 143
risk factors, 107, 114
rna, 101
RNA, vi, vii, viii, xi, 21, 23, 24, 28, 29, 30, 31, 33, 34, 35, 36, 37, 38, 43, 44, 45, 48, 50, 53, 54, 56, 58, 59, 61, 62, 63, 69, 71, 72, 73, 101, 104, 109, 115, 161, 162, 165, 166, 168, 169
RNAi, v, vii, 21, 22, 23, 24, 25, 26, 27, 28, 29, 30, 31, 32, 33, 34, 35, 36, 37, 38
rodent, 146
rodents, 81, 83, 86, 87, 88, 89, 100, 123, 144, 146
Romania, 86, 104, 108, 117, 118
room temperature, 162
Royal Society, 56
rural, 79, 96, 146
Russia, 79, 85, 86, 87, 88, 89, 90, 101, 108, 118, 142, 143, 161
Russian, 85, 87, 89, 90, 92, 97, 98, 99, 139, 140, 141, 142, 143, 145, 161
Russian Academy of Sciences, 161
Rutherford, 147
RXR, 36

S

safety, xi, 171, 174
saline, 62, 131
salivary glands, 25, 26, 27, 104
Salmonella, 135
sample, 50, 62, 69, 71, 73, 110, 130, 162, 167
sampling, 46
Saudi Arabia, 98
Scandinavia, 92
Scanning electron, 17
scavenger, 31
Schmid, 47, 49, 56, 57, 60, 74
sclerosis, 111
search, 107
seasonal pattern, 70
seasonal variations, 57, 75
secrete, 172, 173
secretion, 172, 173
sedimentation, 43
seeds, 113
seizure, 115
seizures, 106, 112
Self, 115
Senegal, 84
sensation, 111
sensitivity, 50, 57, 106, 124, 127, 131
separation, 124, 163, 164, 165, 166
sequencing, xi, 16, 33, 43, 49, 54, 124, 162, 166, 173
serine, 24, 26, 27, 34, 35
serologic test, 91
serologic tests, 91
serology, 147
serum, 91, 111, 112, 114, 172, 173
services, iv, 151
severity, 88, 90, 112, 113, 135, 137, 146
sex, 151
shape, 5, 7, 16, 42
shaping, 24
shares, xi, 27, 171
sheep, 86, 89, 100, 122, 136, 137, 139, 141, 143, 144, 145, 147
shelter, viii, 41
shock, x, 146, 151
shortage, 2
Siberia, 79, 82, 98
sibling, 101
sign, 90, 108
signal peptide, 4
signal transduction, 26, 30
signaling, 26, 30, 32, 168
signaling pathway, 32, 168
signaling pathways, 32
signalling, 36, 38
signals, 27, 29

signs, 89, 90, 92, 106, 108
silk, 29
silkworm, 14, 15, 17, 35, 37, 43, 48, 55, 58
silver, 173
similarity, 4, 5, 6, 25, 26, 27
sinus, 131
siRNA, 24
sites, 9, 20, 58, 139
skin, 60, 113
sle, 81, 83
sleeping sickness, 26
Slovenia, 97
snakes, 81
social behavior, 49
social impacts, 101
social life, 60
sodium, 124
South Africa, 73, 84, 85, 97, 99
South America, v, viii, ix, 59, 61, 69, 72, 78, 79, 82, 83, 98
Southeast Asia, 88, 102
SP, 26, 37, 98
Spain, 87, 88, 99
specialization, 57
species, vii, ix, x, xi, 2, 3, 4, 6, 10, 12, 22, 23, 25, 26, 28, 32, 33, 43, 45, 46, 47, 50, 56, 60, 74, 80, 81, 86, 88, 94, 101, 103, 104, 122, 136, 137, 139, 140, 142, 144, 146, 152, 153, 154, 155, 157, 166, 171
specificity, 16, 48, 174
spectrum, 2, 4, 10, 25, 110, 112
speech, 81, 107, 112, 115
speed, 10, 62, 112
spinal cord, 90, 91, 105, 109, 110, 114
spinal tap, 108, 109
spindle, 5, 12, 14, 16, 17, 20
sporadic, 81, 82, 86, 92
Sri Lanka, 84
St. Louis encephalitis (SLE), 80, 81, 82, 83, 95, 97, 101, 102, 104, 116
stability, 17, 71
stages, 31, 53, 60
Standards, 142
Staphylococcus, 30, 31, 135
Staphylococcus aureus, 30, 31, 135
sterile, 42, 62, 124
steroid, 112
steroids, 112
stimulant, 12, 15
storage, 73

strain, 6, 13, 19, 28, 46, 47, 48, 56, 89, 97, 100
strains, x, 8, 24, 46, 47, 50, 84, 85, 96, 98, 103, 104, 117
strategies, 49, 100, 104
strength, 108, 110, 111, 157
Streptomyces, 6
stress, 36, 166
stroke, 112
structural protein, 9, 101
subcutaneous (SC), 95, 96, 97
subgroups, 43
subjective, 140
sub-Saharan Africa, 26
sugar, 42
sugars, 12
suicidal, 42, 49
suicide, 49
sulfate, 124
summer, 48, 65, 67, 68, 69, 70, 82, 86, 88, 89, 90, 97, 99, 128
Sun, 3, 10
supernatant, 62
suppression, 34, 36, 48, 136, 137
surgical, 98
surveillance, 101, 116, 117
survival, ix, 78, 79, 137
survivors, 82, 90
susceptibility, 16, 27, 31, 34, 107, 129, 136, 157, 160
sustainability, 61
swallowing, 112
Sweden, 85, 97, 139, 140
swelling, 111, 119
Switzerland, 87, 143, 145, 147
symptoms, viii, 50, 59, 60, 61, 63, 64, 65, 66, 67, 68, 69, 70, 71, 72, 86, 89, 90, 105, 111, 113, 136, 137, 138
syndrome, x, 82, 103, 109, 110, 111, 118, 119
synergistic, 8, 10, 15, 16, 19, 20, 122
synergistic effect, 10, 122
synthesis, 14, 20, 24, 25, 28, 39, 58, 167
synthetic, 115

T

taiga, 142
Taiwan, 89, 112
target organs, 113
targets, 15, 32
taxa, 13, 162

taxonomy, 116, 159
technical assistance, 159
temperature, 3, 5, 10, 79, 93, 94, 162
temporal, 114
temporal lobe, 114
tendon, 108, 109
Texas, 82, 83, 102
Thai, 72
Thailand, 83, 89
thalamus, 90, 107
therapeutic, 104
therapeutics, 22
therapy, 80, 107, 112, 115, 116
Thomson, 14
thorax, 51
threatening, 115
threats, 60
thresholds, 57
thrombocytopenia, 105, 136
tick fever, 78, 81, 98
tick-borne disease, 139
tick-borne encephalitis virus, 97, 101, 128, 142
ticks, ix, x, 33, 77, 79, 80, 81, 84, 86, 89, 92, 96, 98, 99, 100, 121, 122, 123, 124, 125, 126, 127, 128, 129, 130, 131, 132, 133, 134, 135, 136, 137, 138, 139, 140, 141, 142, 143, 144, 145, 146, 147
time, ix, 103, 107, 113, 134, 137
time periods, 153
tin, 111
tissue, vii, 1, 2, 51, 104, 107, 113, 114, 162
TNF, 26
tobacco, 13, 28, 29, 36
Tokyo, 41, 51
Toll-like, 27
toxic, 25
toxin, 37, 152
trade, 151
traits, 166
trans, 36, 157
transaminases, 105
transcript, 106
transcriptase, 118, 168
transcription, 28, 53, 62, 73
transcriptional, vii, 21, 23, 26, 30
transcripts, 29, 50
transduction, 26, 30
transfer, 159
transformation, 32, 36
transforming growth factor, 26
transgenic, 13, 16, 27

Transgenic, 33
transgenic plants, 13
transmembrane, 9, 29, 31
transmission, viii, x, 9, 45, 48, 77, 78, 80, 84, 85, 88, 95, 101, 104, 121, 122, 123, 125, 127, 128, 133, 134, 135, 137, 138, 141, 144, 145, 152
transmits, 45, 122
transport, 30, 71, 114
transposons, vii, 21, 23
treatment, 96
trial, 116, 120
tropism, ix, 51, 78, 79, 104, 113
tryptophan, 34
tularemia, 136, 140, 141, 145
tumors, 113
Tunisia, 84
Tuscany, 92
two-dimensional, 173

U

Uganda, 84, 85, 99, 104, 116
UK, 21, 53, 62, 73, 98
ultrastructure, 97
ultraviolet (UV), x, 3, 5, 10, 12, 151, 152, 153, 155, 157, 165
ultraviolet light (UV light), 12, 155, 165
uncertainty, 153
unclassified, 2
unification, 140
unilateral, 111
United States, 46, 56, 57, 60, 71, 81, 95, 104, 108, 117, 119, 122, 142, 144
Urals, 92
urban areas, 79
urinary, 109
urinary retention, 109
urine, 94, 146
Uruguay, v, viii, 59, 61, 63, 64, 65, 66, 67, 68, 69, 70, 71, 72
USDA, 151

V

vaccination, 81, 116
vaccine, 80, 95, 109, 116, 120, 139
vaccines, ix, 78, 80
values, x, 46, 94, 152, 156, 157, 158
variability, 88, 95, 109, 128

variable, 108, 115, 116
variation, 32, 157, 162
vascular, 114
vascular disease, 114
vascular diseases, 114
vector, ix, 36, 37, 48, 77, 79, 92, 96, 100, 102, 103, 116, 122, 135, 137, 138, 140, 142, 144, 145
vectors, ix, 77, 78, 79, 80, 81, 84, 86, 88, 91, 92, 95, 97, 98, 99, 100
velocity, 109
Venezuela, 81, 90
ventilation, 109, 110, 115
vertebrates, vii, 1, 2, 10, 11, 23, 80, 136
vesicle, 30
viral, viii, ix, 77, 78, 79, 80, 84, 85, 90, 91, 96, 100, 102, 104, 106, 107, 108, 111, 113, 114, 115
viral envelope, 6, 20, 104
viral infection, vii, viii, 6, 21, 31, 50, 64, 65, 66, 67, 115, 172, 173
viral meningitis, 108
virology, 50, 52
virulence, 16, 17, 20, 23, 32, 48, 80, 84, 113
virus, vii, viii, ix, xi, 1, 2, 4, 5, 9, 10, 12, 13, 14, 15, 16, 17, 18, 19, 20, 29, 31, 32, 35, 41, 43, 45, 46, 48, 49, 50, 51, 52, 53, 54, 55, 56, 57, 58, 59, 60, 63, 64, 65, 69, 71, 72, 73, 74, 75, 78, 79, 80, 81, 82, 83, 84, 85, 86, 88, 89, 92, 93, 94, 95, 96, 97, 98, 99, 100, 101, 102, 103, 104, 106, 107, 109, 111, 112, 113, 114, 115, 116, 117, 118, 119, 120, 122, 128, 136, 139, 142, 145, 152, 153, 157, 159, 160, 161, 162, 165, 166, 167, 168, 169, 172, 174
virus infection, 16, 19, 32, 53, 73, 74, 75, 79, 94, 100, 101, 102, 107, 109, 117, 118, 119, 120, 139, 168
virus replication, 4, 94, 120, 167
viruses, vii, viii, ix, xi, 1, 2, 3, 5, 6, 7, 9, 10, 11, 12, 13, 15, 17, 19, 21, 23, 24, 31, 32, 35, 36, 41, 43, 45, 46, 47, 48, 49, 50, 52, 53, 54, 55, 56, 57, 58, 59, 60, 61, 63, 64, 65, 66, 67, 68, 69, 70, 71, 72, 73, 75, 77, 78, 79, 80, 82, 84, 86, 88, 89, 90, 93, 95, 96, 98, 101, 104, 106, 109, 111, 112, 115, 116, 120, 152, 159, 160, 162, 166, 168, 169, 171, 172, 173
vision, 111
visual, 111, 112
visual acuity, 111
visual field, 111
voles, 135
vomiting, 90, 105, 108

W

Wales, 96
water, 124, 130, 165
weakness, 108, 110, 111, 112
well-being, 172
West Nile fever, 105, 117, 118, 120
West Nile virus, ix, 78, 79, 81, 84, 86, 95, 96, 97, 99, 102, 103, 117, 118, 119, 120
Western Hemisphere, ix, 78, 79, 103, 117
wheat, 153
wild animals, x, 121, 128
wild type, xi, 5, 55, 161, 166
wildlife, 99
winter, 65, 66, 68, 69, 70
Wisconsin, 139, 144
wisdom, 34
workers, 42, 43, 44, 46, 47, 49, 50, 96, 104
worm, 23, 29
wound healing, 27

Y

yellow fever, 79, 80

Z

zoonosis, 95
zoonotic, ix, 89, 102, 103, 139, 143